LORENZO RAUSA

CNC
50 HOUR PROGRAMMING COURSE

For lathes, ISO Standard functions, Siemens fixed cycles, parametric programming, methods of use

On the cover: images from the programming and graphic simulation software SinuTrain Sinumerik Operate (Siemens A.G.)

Copyright © Lorenzo Rausa 2013
New York (U.S.A.)

info@cncwebschool.com
www.cncwebschool.com

All rights reserved under national law
and international conventions.

This book or parts thereof may not be reproduced, stored in a retrieval system, or transmitted in any form without the permission of the publisher.

First edition: December 2013

ISBN 978-1493713578

*To Alex and Rick,
who truly make the future*

*'With commitment, time and method,
you can achieve all your goals'*

Table of contents

Foreword

1. **Course Introduction (1h) 19**
   1.1 Purpose 19
   1.2 Means 20
   1.3 Method.. 20
   1.4 Duration.. 21
   1.5 Lathe... 21
2. **Start-Up of the Training Software...23**
   2.1 Download of the SinuTrain Operate program............................. 23
   2.2 Installation... 24
   2.3 Creation of the lathe and license activation................................ 25
   2.4 Download of the programs and import into SinuTrain 26
3. **From the Program to Graphic Simulation (2h) 29**
   3.1 Introduction.. 29
   3.2 Opening of the program.. 30
   3.3 Import of tool data 31
   3.4 Graphic definition of the blank part.. 33
   3.5 Start of simulation... 33
   3.6 Program selection before to start the production 36
4. **Name and Direction of the Axes (2h)..37**
   4.1 Layout of the axes according to ISO Standard 37
   4.2 X-axis and the Z-axis.. 38
   4.3 C-axis 39
   4.4 How to determine the positive motion of the rotating axes 41
   4.5 Y-axis... 42
   4.6 B-axis 43
   4.7 A-axis... 44

	4.8	Concept of interpolation	44
	4.9	Programming scheme	45
	4.10	Practical exercise	47
		4.10.1 Movement on the X- and Z-axes and angular orientation	47
		4.10.2 Calculation of the offset values	50
		4.10.3 Duplication, renaming and modification of a program	53

5. Programming Concepts (3h) 55

5.1	Elements constituting a program	55
5.2	Logical programming sequence	56
5.3	Duration of the validity of an instruction	57
	5.3.1 Modal instructions and groups of origin	57
	5.3.2 Self-deleting instruction	59
5.4	Instruction types	59
	5.4.1 Technological instructions	59
	5.4.2 Geometrical instructions	59
	5.4.3 Auxiliary instructions 'M'	60
5.5	Complementary instructions	61
	5.5.1 Entering of comments	61
	5.5.2 Message display	61
5.6	Automatic block numbering	62
5.7	Practical exercise	63
	5.7.1 Program analysis	63
	5.7.2 Automatic block numbering	64
	5.7.3 Deletion of block numbers	66

6. Coordinate Systems (2h) 67

6.1	Machine coordinate system (MCS)	67
	6.1.1 Machine zero point	68
	6.1.2 Characteristic point of the slide	69
6.2	Workpiece coordinate system (WCS)	69
	6.2.1 G54 - G57: part zero point setting	70
	6.2.2 Tool offset	72
6.3	Practical exercise	73
	6.3.1 Setting of the part zero point, use of MDA and JOG	73
	6.3.2 Display of the position in MCS and WCS	75
	6.3.3 Tool offset by touching the workpiece	78

7. Tool Call (2h) 79

7.1	Introduction	79
7.2	T: tool call and function M6	79

7.3 D: tool offset values selection 80
7.4 Correction of the tool wear 83
7.5 Practical exercise 84
    7.5.1 Creation of a tool 84
    7.5.2 Deletion of a tool 86
    7.5.3 Creation of a second tool corrector 87
    7.5.4 Deletion of a second tool corrector 88
    7.5.5 Mounting and removal of the tools in the turret 89
    7.5.6 Saving of tooling data 89

8. Spindle Activation (2h) 91
8.1 Introduction 91
8.2 SETMS: setting of the master spindle 93
8.3 G97: Spindle rotation with constant number of revolutions 94
8.4 G96: setting of constant cutting speed 95
8.5 LIMS=: limitation of the maximum number of revolutions 96
8.6 M3, M4, M5: setting of the rotation direction 97
8.7 Instructions to a spindle which is not the master spindle 97
8.8 Choice of the functions G97, G96 and LIMS 98
8.9 SPOS=: programming of the angular orientation 98
8.10 Practical exercise 99
    8.10.1 Calculation exercises 99
    8.10.2 Creation of a new main program 100
    8.10.3 Creation of a new subprogram 101
    8.10.4 Creation of a new folder 101
    8.10.5 Spindle angular orientation exercises review 102

9. Setting of the Feedrate (1h) 103
9.1 Introduction 103
9.2 G95: feedrate expressed in mm/rev 103
9.3 G94: speed in mm/min 103
9.4 Calculation of the execution time for one pass 104
9.5 Practical exercise 105
    9.5.1 Calculation exercises 105
    9.5.2 Saving of folders and programs 106

10. Absolute and Incremental Coordinates (1h) 107
10.1 G90: absolute programming 107
10.2 G91: incremental programming 109
10.3 Mixed programming 110

    10.4 Diametral or radial meaning of the values associated with X 110
    10.5 Practical exercise.. 111
          10.5.1 Analysis of a program in absolute coordinates 111
          10.5.2 Analysis of a program in incremental coordinates........... 112

11. Basic Functions to Define the Profile (3h) 113
    11.1 G0: rapid movement 113
    11.2 G1: linear interpolation 114
    11.3 G33, G34, G35: threading in multiple passes........................ 115
    11.4 G4: dwell function 116
    11.5 Practical exercise.. 117
          11.5.1 Example of the roughing of a profile 117
          11.5.2 Review of program comprehension 119
          11.5.3 Example of the programming of a threading 120
          11.5.4 Finishing of a profile 123

12. Direct Programming of Rounds, Chamfers and Angles (2h) 125
    12.1 Introduction... 125
    12.2 RND= / RNDM=: execution of a round 125
    12.3 CHR= / CHF=: execution of a chamfer................................ 127
    12.4 FRC= / FRCM: specific feedrate on chamfers and rounds......... 128
    12.5 ANG=: direction of a line defined by an angle...................... 129
    12.6 Practical exercise.. 131
          12.6.1 Point to point and direct programming comparison 131
          12.6.2 Definition of the blank part data 133
          12.6.3 Programming of a workpiece 135

13. Circular Interpolation (1h) 137
    13.1 G2: circular interpolation in clockwise direction 137
    13.2 G3: circular interpolation in counterclockwise direction............. 138
    13.3 I, K, I=AC(…), K=AC(…): progr. of the radius center............... 139
    13.4 Definition of the working plane 141
    13.5 Practical exercise.. 142
          13.5.1 Programming of different radii....................................... 142

14. First Test (2h) 145
    14.1 Introduction to the test 145
    14.2 Tooling operations and cutting parameters 146
    14.3 Drawing of the part to create 147
    14.4 Copying & pasting of program parts..................................... 148
    14.5 Program correction 148

15. **Tool Radius Compensation (1h)** 149
    15.1 Introduction 149
    15.2 G42: Enabling with tool on right side of profile 154
    15.3 G41: Enabling with tool on left side of profile 155
    15.4 Enabling and disabling with G40 156
    15.5 Practical exercise 157
        15.5.1 Program analysis 157
        15.5.2 Test of concept comprehension 160
    15.6 Reloading of complete tool list 162
16. **Advanced Programming Functions (2h)** 163
    16.1 Introduction 163
    16.2 Call of a subprogram 163
    16.3 Repetition of a subprogram 167
    16.4 Labels: references for the setting of skips 168
    16.5 REPEAT: repetition of blocks 169
    16.6 GOTO: skipping parts of a program 171
    16.7 Conclusions 173
    16.8 Practical exercise 174
        16.8.1 Call of a subprogram 174
        16.8.2 Repetition of a subprogram 174
        16.8.3 Use of the REPEAT function 174
        16.8.4 Use of the GOTO function 174
17. **Fixed Cycles (1h)** 175
    17.1 Introduction 175
    17.2 Use of the HELP button 176
    17.3 Insertion of a cycle in a program 178
    17.4 Deletion of a cycle from the program 178
    17.5 Practical exercise 178
        17.5.1 Buttons for cycle management 178
18. **CYCLE62: Profile Selection (1h)** 179
    18.1 Description 179
    18.2 Insertion procedure 179
    18.3 Parameters 180
    18.4 Practical exercise 183
        18.4.1 How to identify a profile in a program 183
19. **CYCLE952: Roughing Cycle (2h)** 185

	19.1	Description 185
	19.2	Insertion procedure 185
	19.3	Parameters 186
	19.4	Practical exercise 190
		19.4.1 External roughing and finishing of a workpiece 190
		19.4.2 Programming alternative with subprogram 192
		19.4.3 Internal roughing and finishing of a workpiece 193
		19.4.4 Use of the tool radius compensation 195

20. CYCLE99: Threading Cycle (1h) 197

    20.1 Description 197
    20.2 Insertion procedure 197
    20.3 Parameters 198
    20.4 Practical exercise 202
        20.4.1 Programming example 202

21. Second Test (2h) 205

    21.1 Introduction to the test 205
    21.2 Tooling operations and cutting parameters 207
    21.3 Drawing of the part to be created 208
    21.4 Program correction and reloading of tool list 208

22. CYCLE930: Cycle for Grooves (1h) 209

    22.1 Description 209
    22.2 Insertion procedure 209
    22.3 Parameters 210
    22.4 Practical exercise 213
        22.4.1 Programming example 213

23. CYCLE82: Drilling Cycle (1h) 215

    23.1 Description 215
    23.2 Insertion procedure 215
    23.3 Parameters 216
    23.4 Practical exercise 218
        23.4.1 Example of the programming of an axial hole 218
        23.4.2 Example of the programming of a radial hole 219
        23.4.3 Modal activation of the cycle with MCALL 220

24. CYCLE83: Deep Hole Drilling Cycle (1h) 223

    24.1 Description 223
    24.2 Insertion procedure 223

	24.3	Parameters 224
	24.4	Parameters relative to the chip breaking 227
	24.5	Parameters concerning the chip removal 228
	24.6	Practical exercise 229
		24.6.1 Execution of a hole with chip break 229
		24.6.2 Execution of a hole with chip removal 230
25.	**CYCLE84/840: Tapping Cycle (1h) 231**
	25.1	Description 231
	25.2	Insertion procedure 231
	25.3	Parameters 232
	25.4	Parameters concerning the chip break 236
	25.5	Parameters concerning the chip removal 236
	25.6	Practical exercise 237
		25.6.1 Execution of a rigid axial tapping 237
		25.6.2 Execution of a compensated radial tapping 239
26.	**CYCLE940: Cycle for Thread Undercuts (1h) 241**
	26.1	Description 241
	26.2	Insertion procedure 241
	26.3	Parameters 242
	26.4	Practical exercise 245
		26.4.1 Execution of an undercut for external thread M16 245
27.	**Use of the C-Axis (4h) 247**
	27.1	Introduction 247
	27.2	C-axis 247
		27.2.1 M70: Activation of the C-axis 248
		27.2.2 Milling of lines on the circumference 250
		27.2.3 TRACYL: cylindrical interpol. on the circumference 254
		27.2.4 TRANSMIT: frontal interpolation 256
	27.3	Practical exercise 259
		27.3.1 Angular orientation and milling on the circumference 259
		27.3.2 Cylindrical interpolation with TRACYL 261
		27.3.3 Frontal interpolation with TRANSMIT 264
28.	**Third Test (2h) 267**
	28.1	Introduction to the test 267
	28.2	Tooling operations and cutting parameters 267
	28.3	Drawing of the part to be created 269
	28.4	Support program for the test 270

- 28.5 Program correction and reloading of tool list 274
29. **CYCLE60: Engraving Cycle (1h) 275**
   - 29.1 Description 275
   - 29.2 Insertion procedure 275
   - 29.3 Parameters (cylindrical interpolation) 276
   - 29.4 Parameters (frontal interpolation) 279
   - 29.5 Practical exercise 282
     - 29.5.1 Example of an engraving on the circumference 282
     - 29.5.2 Example of a front face engraving 284
30. **Parametric Programming (2h) 287**
   - 30.1 Use of parametric programming 287
   - 30.2 Calculation variables 'R' 287
   - 30.3 System variables 289
   - 30.4 Symbols for arithmetic calculations 290
   - 30.5 Symbols for trigonometric calculations 290
     - 30.5.1 Calculation scheme of the trigonometric functions 291
   - 30.6 Result management instructions 292
   - 30.7 Practical exercise 293
     - 30.7.1 Calculation test in operating mode MDA 293
     - 30.7.2 Programming of a workpiece family 294
31. **Conditional Program Skips (2h) 299**
   - 31.1 The algorithm 299
   - 31.2 Comparison operators 299
   - 31.3 Logical operators 300
   - 31.4 STOPRE: stop function for the program execution 301
   - 31.5 Practical exercise 302
     - 31.5.1 Creation of the algorithm 302
     - 31.5.2 Parametric counter with WHILE 304
     - 31.5.3 Parametric counter with IF 305
     - 31.5.4 Algorithm for the execution of 36 longitudinal holes 306
32. **Three-Axis Mill: Programming Basics 309**
   - 32.1 Introduction 309
   - 32.2 Layout of the axes in a vertical mill 309
   - 32.3 Machine zero point and definition of the part zero point 310
   - 32.4 TRANS/ATRANS: incremental shift of the part zero point 312
   - 32.5 Position of the point controlled by the NC and tool geometry 313

- 32.6 Setting of tool rotation and feedrate 315
- 32.7 Practical exercise 316
  - 32.7.1 Introduction 316
  - 32.7.2 Creation of a three-axis mill (X, Y, Z) 316
  - 32.7.3 Download of the programs and import into SinuTrain 317
  - 32.7.4 Direct selection of the tools in the program 318
  - 32.7.5 Graphic definition of the blank part 320
  - 32.7.6 Drawing of the part to be created 324
  - 32.7.7 Program, phase 1: execution of the external profile 325
  - 32.7.8 Program, phase 2: roughing of the internal profile 326
  - 32.7.9 Program, phase 3: finishing of the internal profile 327
  - 32.7.10 Program, phase 4: execution of the holes 328
  - 32.7.11 Program, phase 5: activation of graphic simulation 329
  - 32.7.12 Programming example with the use of TRANS 330

33. Translation of technical terms 333
- 33.1 Machine components 333
- 33.2 Programming 334
- 33.3 Tooling operations 336
- 33.4 Materials and related terms 337
- 33.5 Notes 338

Conclusion

References of the figures

14

Foreword

This book was written for the purpose of meeting the needs of all people looking for a programming course based on ISO Standard language, with a special focus on numerically controlled lathes and in combination with a software able to reproduce a real NC on the computer and to perform the graphic simulation of the program created.
The goal is to offer a simple and comprehensive guide allowing the integration of theoretical learning with practical work experience on a machine tool.
As the title suggests, the course is subdivided into 50 hours. The license for the free use of the training and graphic simulation software, which may be downloaded from the Internet according to the instructions provided in this book, has a validity of sixty days.
The topics covered range from basic programming information to more advanced sessions; the theoretical lessons are combined with practical exercises, which stimulate the learning process and avoid the course being reduced to a merely passive reading of notions and concepts.
The basic prerequisites for the reader are the ability to use a computer, the understanding of mechanical drawings and the knowledge of basic mechanical concepts and of the tooling operations on machine tools.
The total number of hours necessary for the completion of the theoretical and practical part will always be specified at the beginning of each chapter. This will allow both the student and the teacher to select the topics to be covered based on available time and to assess progress achieved by completion of the exercises within set timeframes.
The machine that is being examined is a three-axis lathe (X, Z, C) with a turret able to support fixed or driven tools and operated by a Siemens Sinumerik Operate numeric control, Series 840D.
The type of lathe and of numeric control have been selected so as to base the learning process on modern products that are widely available around the world.

The SinuTrain training software is also a Siemens product and faithfully replicates all the functions and pages of a numeric control on a computer screen.
The topics are explained in a language that is simple and easy to understand, further enhanced with over 220 images and 50 programming examples.
The training and graphic simulation software, all the programs for lathes and mills used in this course and all the images contained in this book, which may be printed or viewed in order to simplify the analysis of technical drawings or course explanation in the classroom, maybe downloaded from the website: cncwebschool.com.

At the end of this course, the concepts applied to the programming of the lathe will be taken up again in a chapter that allows an easy understanding of the programming principles of a three-axis vertical mill (X, Y, Z).

Foreign languages have not been forgotten either: the last chapter contains a list of the more widely used technical terms with their respective translation from English to Italian and German.

I would like to thank the Don Bosco Technical Institute of Milan and all its teachers.

A special thanks to Randy Pearson (who works with Siemens as International Business Development) for his valuable contribution to review the English translation and his precious technical suggestions.

I would also like to thank SIEMENS, DMG MORI and SANDVIK COROMANT for their constant support and dedication during the drafting of this book and all my colleagues from GILDEMEISTER ITALIANA.

LORENZO RAUSA

1. Course Introduction (1h)
(Theory: 1h)

1.1 Purpose

The specific purpose of this course is to learn how to create the complete program of a part starting with its drawing. However, the course consists in the achievement of many small goals.

The correct programming of a numerically controlled lathe is the result from the combination of different types of knowledge; these are the small goals represented by the conclusion of every chapter.

Learning the meaning of codes and functions is not sufficient for the creation of a part, because it is also necessary to understand the types of operations that the machine is able to perform; it is furthermore important to set the tools and to calculate the cutting parameters, all useful steps in the creation of the sequence of the tooling operations to be programmed, i.e. the execution of the working cycle.

After completion of the part-program, it is necessary to know how to use the control panel of the machine, the operating sequences to enter the program, to edit it, to save it and to reload it, to offset the tools, to view the graphic simulation, test the program and start the machine in automatic cycle.

These goals can be achieved through commitment and allocation of study time, supported by the learning method provided in this book.

Last but not least, those of you who use foreign languages in the field of machine tools should learn by heart at least half of the technical terms translated in the last chapter.

1.2 Means

This handbook associated to the training and graphic simulation software (SinuTrain Sinumerik Operate) provides the means to achieve the set-out goals.

Sinumerik Operate is the name of the latest operating system created by Siemens for the operation of its numeric controls.

The SinuTrain program faithfully replicates all the functions and screens of a numeric control on a computer screen, which, during this course, becomes an actual NC lathe.

Furthermore, a keyboard reproducing the control panel may be connected to the computer. Its use makes the simulated experience ever more similar to the programming of a real numerically controlled machine.

Fig. 1. Optional control panel for SinuTrain

1.3 Method

This course comprises chapters and paragraphs for both theoretical and practical learning. Paragraphs on theory contain drawings and diagrams that simplify the understanding of the text. The collection of all images and programs used during this course may be downloaded from www.cncwebschool.com in the DOWNLOAD section. The first practical experiences consist in the utilization of pre-drafted programs, which are useful to the participant's initial understanding of the numeric control and its potential.

Later you will learn how to write new programs with difficulty levels that are commensurate to the acquired experience.

During the practical exercises, the reader is constantly guided by the respective operating procedures; in the learning phase on the use of a machine tool, the sequence of the buttons to push is always the most complex part to remember and the most boring part to write down.

The learning method has been developed so that even beginners may complete the course and understand all the most complex functions and programming methods.

Periodical tests are offered in order to help the students and teachers assess progress achieved or to highlight the topics for review.

The course is based on the understanding of the 'ISO Standard' functions, i.e. the programming language at the basis of all numeric controls, whose commands have been encoded and, thereby, standardized by the International Organization for Standardization.

Knowledge of the ISO Code enables the programmer to operate on various numeric controls reducing the inconveniences caused by their inevitable differences.

1.4 Duration

This is a fifty-hour course. The total number of hours necessary for the understanding of the theoretical part and for carrying out the practical exercises will always be specified at the beginning of each chapter.

The license for the free use of the SinuTrain program has a validity of sixty days, which is enough time to guarantee the completion of the course and the consolidation of the acquired knowledge.

1.5 Lathe

The machine that is being examined is a three-axis lathe (X, Z, C) with driven tools, it is a single-channel machine and therefore it only needs one program to control the movement of all the axes.

This is the configuration of most lathes in workshops all over the world and it is the ideal starting point to understand the operation and the programming of more complex lathes equipped with more axes, more spindles or more channels.

The most important parts of the lathe are: the spindle, the turret and the numeric control.

The spindle holds the workpiece between three jaws that are normally powered by an oil hydraulic circuit. The spindle, also referred to as chuck, is able to place the center of the workpiece on its rotating axis.

The turret simulated in the training program offers twenty available positions which can all be equipped with fixed or driven tools.

The NC controlling the lathe is produced by Siemens, Series 840D. This numeric control may be programmed in ISO Standard language or via ShopTurn, a conversational program designed for lathes.

The goal of this course is to teach ISO language using the features made available by ShopTurn for the setting of the tools and for the execution of the graphic simulation.

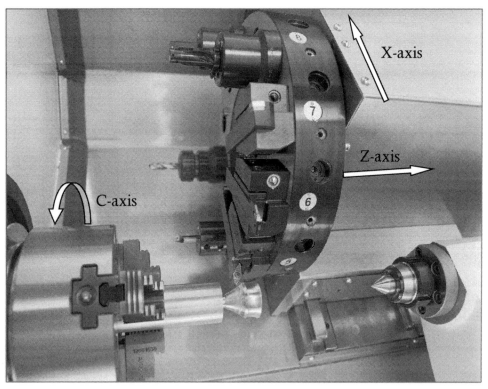

Fig. 2.   Example of a lathe with 3 axes and driven tools

2. Start-Up of the Training Software

2.1 Download of the SinuTrain Operate program

In order to proceed with the next step the computer needs to be connected to the Internet.

Minimum PC requirements necessary for the installation and correct functioning of the SinuTrain program:

Hardware:	Processor: 2 GHz or higher, min. RAM: 1 GB, USB data port
Disc Capacity:	Availability of approx. 3 GB for complete installation
Operating System:	Windows XP Professional (Service Pack 3), Windows 7 Home Premium, Professional, Ultimate, Enterprise (32- and 64-bit)
User Settings:	PC administrator rights required for installation and use
License:	The license is activated during its first installation. It can be used only once and will expire after 60 days.

Fig. 3. Minimum PC requirements

Visit the website www.cncwebschool.com and access the DOWNLOAD area, click on SOFTWARE; the Siemens website opens. Register and write down the access data you created to have them at your disposal for future access:

Username:	Password:
....	....

Fig. 4. Personal access data to Siemens website

After the 'login', activate the link for the download of the latest version of the simulation and training program named:

SinuTrain SINUMERIK Operate 4.4 Ed.3

A window for the *Download of the file* opens asking you if you want to save or open the compressed folder.

Choose to save the folder and wait for the completion of the download.

Close the navigation window, select the downloaded folder with the pointer, push the right button of your mouse and select: *Extract all, Next, Next.*
The installation folder is now ready to be used.

2.2 Installation
Open the folder and start the installation process.
Make sure to select English when asked for the language settings of the numeric control.
On the next screen, three different programs are offered for installation.

Automation License Manager is the program for the management of the licenses bought from Siemens. It is not necessary for the course but could be useful in the future (optional choice).

SinuTrain is the name of the training program for the course (please select).

HMI Cad Reader is an application which allows to open drawings in electronic format (with .dxf extension), to graphically select a profile or the position of a hole and to convert them into a program or subprogram expressed in ISO codes (optional choice).

Wait for the completion of the installation of the selected programs.

2.3 Creation of the lathe and license activation

Start the program with the *SinuTrain SINUMERIK Operate* icon that has been created on the desktop of your PC, then push START.

SinuTrain starts with an empty window which in the future will show the list of all machines created in it.

Now proceed with the creation of the lathe that you want to use during the course.

Push the NEW icon in the upper part of the screen, leave the first point *Create a new machine configuration from a template* selected and push NEXT.

This window shows you a list of standard machine templates preconfigured in SinuTrain. Select *Lathe with driven tool* and press NEXT.

Now choose the name of the lathe, describe its basic features, set the language you want to use and the dimensions of the window which reproduces the machine video. Enter the following information:

GENERAL	*Machine name:*
	LATHE: 50 Hour Programming Course
	Description:
	SP1-spindle (main spindle),
	X-axis (linear geometry axis),
	Z-axis (linear geometry axis),
	SP3-spindle (driven tool)
LANGUAGE	English - English
RESOLUTION	640x480

Push FINISH. The machine has been created and is now displayed on the starting page of the program.

Select the newly created lathe and push the START icon.

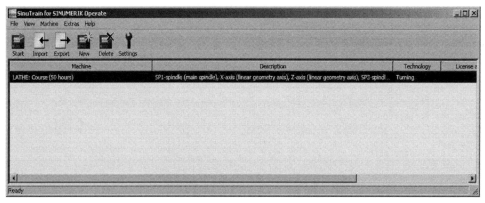

Fig. 5.   Starting window of the simulation program

A window appears saying that no valid License Key was found.

Select the *SinuTrain Operate (Trial)* License Key with your mouse, then push ACTIVATE, then OK to the message shown subsequently.

2.4   Download of the programs and import into SinuTrain

Open the website www.cncwebschool.com and access the DOWNLOAD area, click on LATHE_3_AXES and download the folder T3_PROG containing all the programs used during the course.
Select the downloaded compressed folder with the pointer, push the right button of your mouse and select: *Extract all, Next, Next.*

Now import the programs into the training software. Like a real machine tool, SinuTrain is able to upload and download data to and from an external memory connected via USB port.

Close the SinuTrain program selecting *File* and then *Exit*.

Copy the folder T3_PROG onto an empty USB memory stick.

Open SinuTrain again, select the newly created lathe and push START from the icons in the upper part of the screen.
On the control panel, click PROGRAM MANAGER.

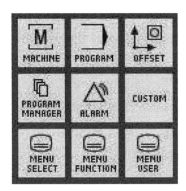

Fig. 6. Buttons for the selection of the operating environments

After selecting USB from the horizontal softkeys, the content of the USB memory is displayed on the screen.

Select the folder T3_PROG with the arrows and push the yellow INPUT button to open it.

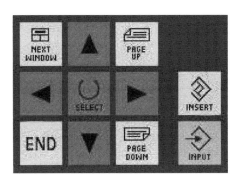

Fig. 7. Buttons for cursor movement and data entry

Now, place the orange selection bar on the first folder in T3_PROG named 01_EXERCISES.
Press MARK from the vertical softkeys and move down with the arrows until the whole content of the folder has been selected.

Press COPY.

Push NC from the horizontal softkeys.

Select the folder WORKPIECES with the arrows and push PASTE from the vertical softkeys.

Now all the programs and the file containing the tooling data are ready for use.

3. From the Program to Graphic Simulation (2h)
(Practice: 2h)

3.1 Introduction

This exercise enables you to quickly understand all the contents of the course and consists in beginning from a program that has already been written in order to obtain a graphic simulation of the tooling operations and the finished solid of the part to be produced.

Furthermore, all the procedures for opening the program, defining the dimensions and the shape of the blank part, importing the tool data, starting the simulation and using the display options for the workpiece will be specified.

The part that you are about to create contains many of the tooling operations the machine is able to perform: roughing and finishing of the external profile, forming of the thread undercut, external threading with multiple passes, center drilling, axial drilling and tapping, radial drilling with angular orientation of the spindle, milling of the square with side of 44 mm by interpolation of X-C and engraving of the inscription 'CNC' on the circumference.

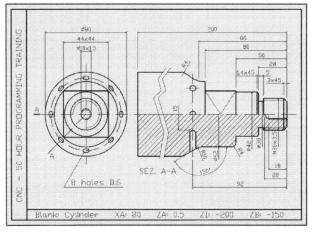

Fig. 8. Technical drawing of the introductory workpiece of the course

3.2 Opening of the program

It should be remembered to follow this procedure every time when it is necessary to open a program in order to modify it or to execute the graphic simulation. Use the mouse to push the buttons displayed on the screen.

The control panel allows the operator to reach the program page either by pushing MENU SELECT and then PROGRAM MANAGER as shown in the following figure or by pushing PROGRAM MANAGER directly from the keypad section for the operating environments.

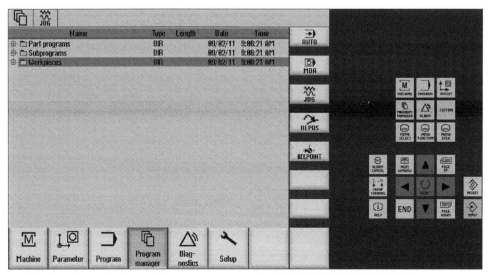

Fig. 9. Organization of the programs on the PROGRAM MANAGER page

Program Manager is the operating environment allowing you to manage and display the programs contained in the NC's memory.

The *Part programs* folder contains the list of main programs created for the production of the parts with the extension .MPF (Main Program File). This folder may not contain other subfolders.

The folder *Subprograms* contains the list of the programs that can be called by the main programs; they have the extension .SPF (Sub Program File). They are secondary programs used for streamlining and simplifying the reading of the main programs.

The folder *Workpieces* may contain other subfolders and offers the possibility to organize the programs in different subgroups or to group all the programs for the production of a certain workpiece in one single folder, accordingly called 'workpiece folder'.
With the arrows, select the WORKPIECES folder, open it by pushing the yellow INPUT button, select the folder CHAP_03 and open it with INPUT.
In the folder you will find the program to be used in this chapter.
In order to open it select PRG_03_01 with the arrows and push INPUT.

3.3 Import of tool data

The data of the tools used by the program PRG_03_01 are gathered in a file that needs to be imported before starting the graphic simulation. Push PROGRAM MANAGER and then open the folder 01_EXERCISES, select the file TOOL_LIST with the arrows and push the yellow INPUT button. The NC recognizes that you want to load the tool data and proposes the following dialog box.

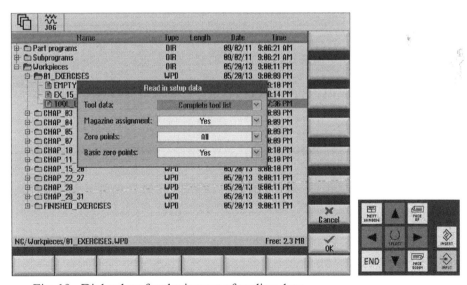

Fig. 10. Dialog box for the import of tooling data

Tool data: with the drop-down menu or the SELECT button placed at the center of the arrows, select: *Complete tool list*. This option overwrites the complete list of tools already defined in the machine.

If you select *No*, this means that you do not want to load the tool data but only the zero points and the basic zero points.
Move to the next items with the arrows.

Magazine assignment: with the drop-down menu or the SELECT button placed at the center of the arrows, select: *Yes*. This option loads all the tools into the same positions of the magazine where they were located when they were last saved. By selecting *No* the tools will be loaded into the positions following the 20 positions available in the magazine, allowing their future relocation on the turret using the LOAD and UNLOAD buttons.

Zero points: with the drop-down menu or the SELECT button placed at the center of the arrows, select: *All*. It is possible to upload the zero points of the axes. Select *No* if you do not want to load these data.

Basic zero points: with the drop-down menu or the SELECT button placed at the center of the arrows, select: *Yes*. This option allows to load not only the axes zero points shift, but also the basic zero points shift.
Then press OK and confirm again with OK your intention to overwrite the current data.

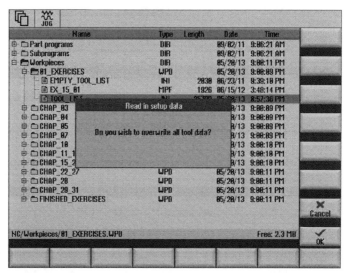

Fig. 11. Window for the confirmation and overwriting of the tool data

3.4 Graphic definition of the blank part

In order to obtain a graphic simulation faithfully replicating reality it is necessary to define the shape and the dimensions of the blank part to be worked on. In all the technical drawings analyzed during this course the data of the blank part are specified in the lower part of the drawing. They have the following meaning:

Blank part:	Shape of the blank part (e.g. cylinder)
XA:	External diameter of the blank part (e.g. 80 mm).
ZA:	Value of the machining allowance on the front face of the blank part (e.g. 0.5 mm).
ZI:	Length of the blank part. If by pushing SELECT you select ABSOLUTE (recommended), the length refers to the part zero point, if INCREMENTAL, the length refers to the front face of the part, machining allowance included.
ZB:	Extension of the face of the blank part from the jaws of the chuck. For the selection of absolute or incremental the same applies as for ZI.

Fig. 12. Description of the blank part dimensions

We will see later on how to enter this information into the top of the program in the block: *WORKPIECE(,,,"CYLINDER",0,0.5,-200,-150,80)*.

3.5 Start of simulation

Following the procedure described in paragraph 3.2, open the program PRG_03_01 with INPUT and press the horizontal icon in the lower right corner: SIMULATION (repeat this step two times at first program start).

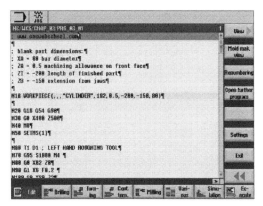

Fig. 13. Program opened and ready for simulation

Now use all the different display options:
SIDE VIEW, 3D VIEW, open FURTHER VIEWS to explore FACE VIEW and HALF CUT VIEW.

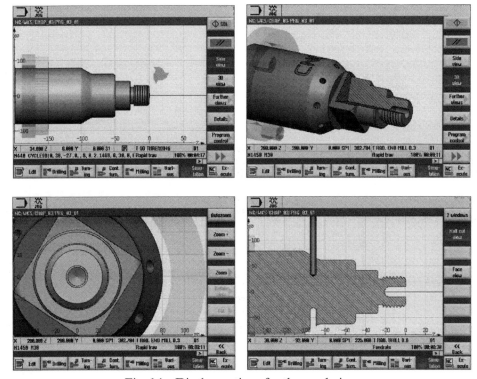

Fig. 14. Display options for the workpiece

Now discover, for each of the display modes, the additional features offered by the DETAILS icon in relation to ZOOM and CUTTING operations of the workpiece.

Then push PROGRAM CONTROL to use the icons which allow to change the execution speed of the profile through potentiometer management: OVERRIDE+, OVERRIDE-, 100% OVERRIDE.

After selecting 3D VIEW, PROGRAM CONTROL, push the OVERRIDE- icon and reduce the feedrate to 80%, then activate the SINGLE BLOCK icon to execute the program line by line.

This setting is very useful for the operator to associate the tool movements with the functions entered into the program, or to find the blocks causing an error in the execution of the profile.
Push BACK and start the execution of the program by pushing the green icon once and again, SBL stands for Single Block, which means that the single block mode is enabled.

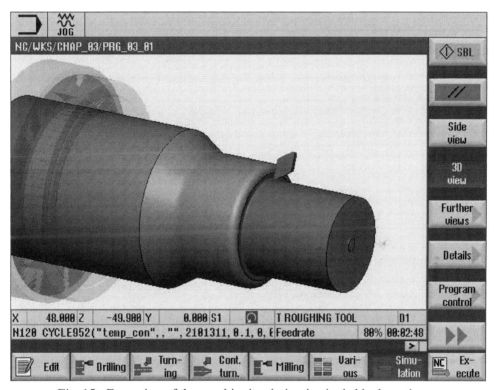

Fig. 15. Execution of the graphic simulation in single block mode

To exit the simulation and return to the program editor use the EDIT icon in the lower left corner of the NC screen.
Review the topics covered in this chapter and practice the use of the graphic simulation until you feel sufficiently confident.

3.6 Program selection before to start the production

Before to start the production of a workpiece it is necessary to check the program by means of the graphic simulation, mount the tools on the turret, set the tool offset and then produce the first parts.

In order to determine which of the programs in the NC shall be carried out by pushing CYCLE START it is necessary to perform a "program selection".

Please remember that this is an essential procedure to follow in order to start the production of the machine, but it is not necessary during the execution of this course.

Push PROGRAM MANAGER, select the file PRG_03_01 with the arrows, then push the first icon on top: EXECUTE.

The NC prepares for the start of the automatic cycle setting the AUTO operating mode and displaying the current position of the axes.

Enable the spindle rotation and the feed by pushing the green buttons SPINDLE START and FEED START. They were successfully enabled when the green light above the buttons is on.

Use the mouse to rotate the potentiometer of the spindles and the feed to 100% as shown in the following figure.

After pushing the green button CYCLE START, you begin to see the execution of the selected program on the screen. Now the lathe is producing the part in automatic cycle.

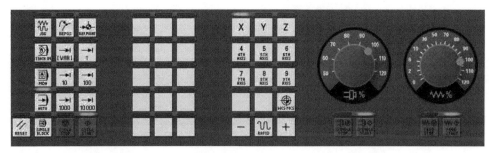

Fig. 16. Activation and setting of the potentiometers for the execution of the program in automatic cycle

4. Name and Direction of the Axes (2h)
(Theory: 1h, Practice: 1h)

4.1 Layout of the axes according to ISO Standard
Each axis is defined by the moving direction of the slide and is characterized by the capacity to interpolate with other axes in the machine. The presence of multiple axes in the machine means that the slide or slides move in various directions. ISO Standards have determined the name of every axis according to its direction and have defined their positive sign according to the sketch in the following figure:

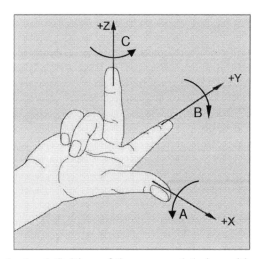

Fig. 17. Right-hand rule: definition of the axes and their positive direction according to ISO Standards. The positive direction is always determined in relation to the moving path of the tool on the workpiece.

This rule is called the "right-hand rule". The thumb represents the X-axis, the index finger the Y-axis and the middle finger the Z-axis. The same standard furthermore defines the names of the rotating axes. The axis rotating around X is called A, the axis rotating around Y is called B and the axis rotating around Z is called C.

4.2 X-axis and the Z-axis

In a lathe, the main axes are the X-axis and the Z-axis. They define the working plane X-Z. The movement of the tool in this plane together with the simultaneous rotation of the workpiece around the Z-axis leads to the creation of a solid by revolution. The Z-axis is therefore considered the axis which generates the part.

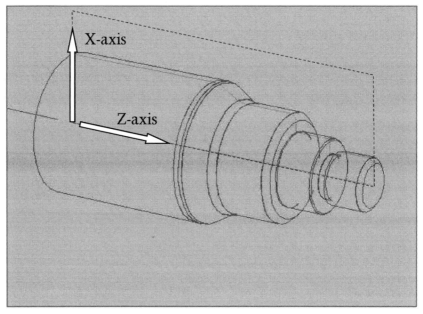

Fig. 18. Solid of revolution around the Z-axis of the profile described on the plane X-Z

The X-axis is the transversal axis of the diameters. The values programmed on the X-axis define the diametrical position of the tool with respect to the rotating axis of the workpiece, which is always considered equal to zero. The distance traveled by the tool, from one diametrical value to the other, corresponds to half of the difference between the two values; it is necessary to use this information during the programming of chamfers or grooves, which are often specified on the drawings with radial values.

The Z-axis is the longitudinal axis of the lengths. All values programmed on the Z-axis are real and refer to the part zero point which is normally located on the front face of the workpiece. The difference between two values in Z corresponds to the real distance traveled by the tool.

4.3 C-axis

The numeric control, thanks to its calculation functions, furthermore offers the opportunity to use the rotating axis of the spindle as an interpolating axis, i.e. it is capable of coordinating its movements on the basis of the movements of the other axes. The rotating axis of the spindle is always coaxial to the Z-axis and is therefore called the C-axis. With the C-axis, it is possible to perform milling and drilling operations, and its use is always associated with the presence of rotating tools in the machine which are called driven tools or live tools.

The C-axis can be used to angular orientate the spindle in the way to perform radial holes along the X-axis direction or out-of-center holes along the Z-axis direction.

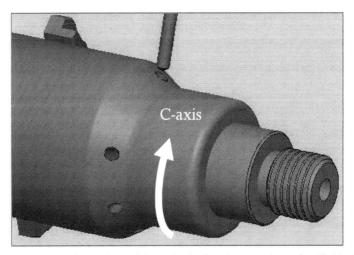

Fig. 19. Angular orientation of the spindle for the creation of radial holes

In lathes designed for processing very large workpieces, the stationing of the spindle in a certain angular position is guaranteed by the presence of a mechanical brake acting directly on a disc combined with the spindle itself.
In smaller machines the absence of the mechanical brake shows that the angular orientation and therefore the blocking of the spindle is obtained by keeping the motor of the spindle electrically active. The motor torque is the power used to contrast every movement caused by the tooling operations performed on the workpiece.

Another way of using the C-axis is to interpolate C with Z to perform milling operations on the workpiece surface. The programmed profile is described on a cylinder, and this is why this type of interpolation is called: cylindrical interpolation.

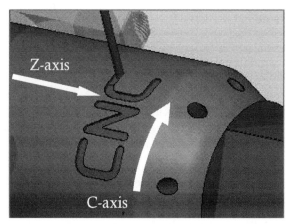

Fig. 20. Example of a cylindrical interpolation C-Z

It is furthermore possible to interpolate the C-axis with the X-axis. This option allows the execution of out-of-axis milling operations on the frontal plane of the workpiece using driven tools coaxial to the Z-axis. In this case the numeric control is able to transform the C-axis into a virtual Y-axis, i.e. an axis which is transversal to the workpiece and which allows for the creation of profiles of any type described on the frontal plane of the workpiece.

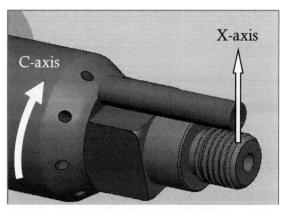

Fig. 21. Example of a frontal interpolation C-X

4.4 How to determine the positive motion of the rotating axes

The positive motion of the rotating axes is determined on the basis of the rule of the closing direction of the right hand.

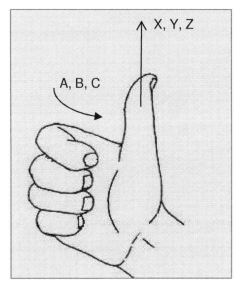

Fig. 22. Right-hand rule to determine the positive motion of the rotating axes

**This rule determines the positive motion of the rotating axis, it is always refered to the movement of the tool.
If, on the contrary, the axis moves workpiece, as in the case of the C-axis, the real movement of the spindle is in the opposite direction.**

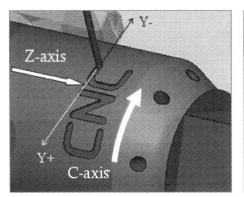 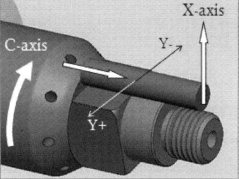

Fig. 23. C-axis positive programming direction and real movement of the workpiece

4.5 Y-axis

The Y-axis is present when the manufacturer of the lathe wants to offer a wider range of milling operations that the machine is able to perform. This axis is linked to the use of driven tools. Through the real Y-axis (and not virtual like the C-axis), it is possible to perform milling operations with flat bottom using radial driven tools. In this case the profile lies on the plane Y-Z. Just think about how a key is created and subsequently enlarged, the presence of a real Y-axis is the only way to create it, making the concept of a lathe ever more similar to that of a mill.

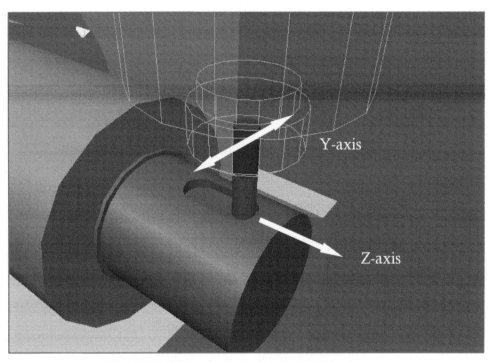

Fig. 24. Milling of a key using the real Y-axis

4.6 B-axis

The B-axis is the axis which rotates around the Y-axis; in lathes, it is used to perform inclined drilling and milling operations or gear hobbing with inclination with respect to the Z-axis. In this case, as for the Y-axis, the B-axis gives the lathe a higher flexibility of use for the milling operations.

The presence of the B-axis in lathes is restricted to a limited series of machines, usually equipped with an automatic tool magazine with conical attachments instead of the more traditional rotating turrets.

The tool magazine, considered as a system for the deposit of tools outside of the working area, is usually a carousel or a chain type and allows for a much higher number of tools. It is therefore better suited to satisfy the complex tooling operations that this type of machine is able to perform.

The presence of a B-axis indicates that we are dealing with a machine which can not be simply defined as lathe, but is in fact a universal machine able to perform turning operations as well as the most complex milling operations.

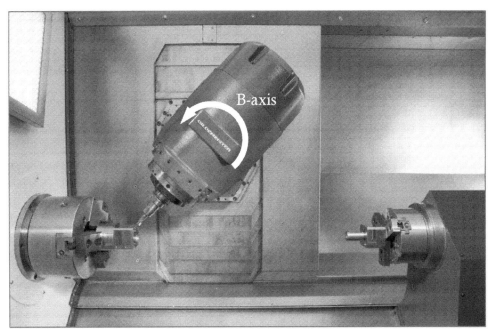

Fig. 25. Universal lathe with B-axis

4.7 A-axis

The A-axis is the axis rotating around X. The configuration of the majority of lathes does not include an A-axis as its functions are the same as the ones offered by a real Y-axis.

On the image below you see a machine with an A-axis which is able to perform rectilinear milling operations on the plane Y-Z thanks to the interpolation of the A- and Z-axes.

Fig. 26. Lathe with A-axis

4.8 Concept of interpolation

We have seen how the tool moves on independent Cartesian axes.

Interpolation means the coordinated movement of one or more axes according to a precise geometrical logic, performed with a specific speed.

Inclined lines, like those programmed to define a conical shape or a chamfer, are executed by interpolation of the X-axis with the Z-axis. In this case, the numeric control coordinates the simultaneous movement of the two axes so that the tool path is a straight line.

Milling operation on a cylindrical surface is obtained by interpolation of the C-axis with the Z-axis.

Every time you program a working movement, you'll have an interpolation; **the geometrical logic is defined by the current function enabled or set in the block, speed is defined by the programmed feed rate.**

4.9 Programming scheme

Most of the numerically controlled lathes use the programming scheme shown in the following figure.

The values associated with X are the diameter on which the tool or slide is located. Most of the time they are positive, X0 is on the rotating axis of the workpiece, negative values represent a position under the rotation center of the workpiece.

The values associated with Z express the longitudinal position of the tool or slide, Z0 is almost always defined on the front face of the workpiece, positive values mean that the position is outside of the workpiece, negative values mean that the tool is in operating phase or that it is in any case positioned beyond the front face of the workpiece.

The figure below shows the programming scheme which needs to be used for the assessment of the positive direction of the axes, in order to determine the clockwise and counterclockwise rotation of the circle arcs included in the profile of the workpiece, to define the right and left position of the tool with regard to the cutting direction and to assess the angle to be programmed in the event of inclined lines.

Standard ISO 841 defines an axes coordinate system (fig. 17) where the axes positive direction is always referred to the movement of the tool.

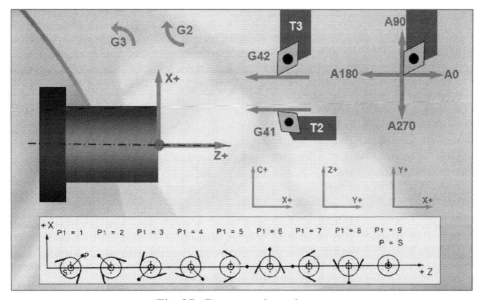

Fig. 27. Programming scheme

The mechanical configuration of the analyzed lathe is the most widely used configuration and includes a turret moving on the workpiece which is fixed and held by the spindle.

Fig. 28. Traditional lathe where tool moves on the workpiece

In some machines the real movement of the Z-axis or of the X-axis is made by the workpiece while the tool remains still. These features, due to technical choices of the manufacturer, normally don't change the programming scheme.

Fig. 29. Lathe with real movement of the Z-axis and of the X-axis on the workpiece

4.10 Practical exercise

4.10.1 Movement on the X- and Z-axes and angular orientation

This exercise aims at clarifying the meaning of the movement values to be programmed on the X- and Z-axis and of the angular values in relation to the orientation of the spindle.

A simple program containing movements on two axes is used to begin with. The executed operations are: turning along the Z-axis, two chamfers by interpolation of the Z-axis with the X-axis, four radial holes at 90 degrees obtained by angular orientation of the spindle.

The dimensions of the blank part are already in the program and the tools are those already used for the execution of the preceding workpiece.

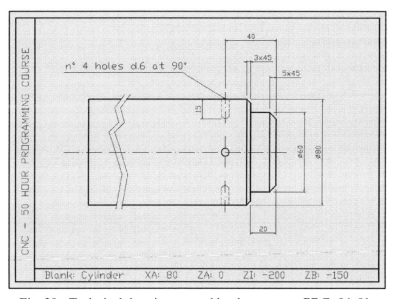

Fig. 30. Technical drawing created by the program PRG_04_01

Open the folder CHAP_04 and the program PRG_04_01 following the procedures specified in chapter 3.

Start the simulation by using the SIDE VIEW option and set the execution of the program to SINGLE BLOCK.

Below you will find the program for the creation of this part, try to proceed with the execution of the program while at the same time carefully reading the comments.

Check the correspondence between the drawing, the programmed value and the movement of the axis. It is not necessary to understand the whole program for this exercise.

```
N10 ; blank part dimensions:
N20 ; XA = 80 bar diameter
N30 ; ZA = 0 machining allowance on front face
N40 ; ZI = -200 length of finished part
N50 ; ZB = -150 extension from jaws
N60
N70 WORKPIECE(,,,"CYLINDER",0,0,-200,-150,80)

N80 G18 G54 G90
N90 G0 X400 Z500
N100 M8
N110 SETMS(1)

N120 T1 D1 ; LEFT HAND ROUGHING TOOL
N130 G95 S1800 M4
N140
N150 ; COMPREHENSION OF THE MOVEMENTS OF THE X AND Z AXES
N160
N170 ; DIAMETRAL MOVEMENT OF 20MM FROM X80 TO X60 MM
N180 G0 X80 Z0
N190 G0 X60
N200 ; THE REAL MOVEMENT OF THE TOOL IS 10 MM
N210
N220 ; LONGITUDINAL MOVEMENT OF 20MM FROM Z0 TO Z-20
N230 G1 Z-20 F0.2
N240 ; THE REAL MOVEMENT IS 20MM
N250
N260 G0 X62 Z0
N270
N280 ; FOR THE CREATION OF A CHAMFER 5x45
N290 ; THE MOVEMENT ON THE X-AXIS
N300 ; IS DOUBLE COMPARED TO THE MOVEMENT ON THE Z-AXIS
N310 G0 X50 ; INITIAL DIAMETER OF THE CHAMFER
N320 G1 X60 Z-5 ; FINAL VALUE IN X AND Z
N330
N340 G1 Z-20 ; LENGTH OF THE TURNING
N350 ; IF THE SECOND CHAMFER IS 3x45, THE INITIAL COORDINATE
N360 ; OF X IS 6MM BEFORE THE FINAL DIAMETER
N370 G1 X74 ; INITIAL DIAMETER
N380 G1 X80 Z-23 ; ARRIVAL POINT IN X AND Z OF THE CHAMFER
N390
N400 G0 X200
```

```
N410 G0 Z200
N420
N430 ; COMPREHENSION OF THE VALUES FOR THE ANGULAR
ORIENTATION OF THE SPINDLE
N440
N450 ; CREATION OF 4 HOLES, STAGGERED BY 90 DEGREES
N460 ; POSITION OF THE FIRST HOLE AT ZERO DEGREES
N470 ; ANGULAR ORIENTATION OF THE SPINDLE AT 0 DEGREES
N480 SPOS=0
N490 SETMS(3)
N500
N510 T8 D1; RIGHT HAND RADIAL DRILL D.6
N520 G95 S1200 M3
N530 G0 Z-40 ; LONGITUDINAL POSITION OF THE HOLES
N540 STR_HOLE1:
N550 G0 X84
N560 G1 X50 F0.1 ; FINAL DIAMETER OF THE DRILL
N570 G4 S2
N580 G0 X84
N590 END_HOLE1:
N600
N610 ; ANGULAR POSITION OF THE SECOND HOLE
N620 SPOS[1]=90
N630 REPEAT STR_HOLE1 END_HOLE1
N640
N650 ; ANGULAR POSITION OF THE THIRD HOLE
N660 SPOS[1]=180
N670 REPEAT STR_HOLE1 END_HOLE1
N680
N690 ; ANGULAR POSITION OF THE FOURTH HOLE
N700 SPOS[1]=270
N710 REPEAT STR_HOLE1 END_HOLE1
N720
N730 G0 X200
N740 G0 Z200
N750
N760 M30
```

4.10.2 Calculation of the offset values

This exercise assesses your understanding of the calculation method used to define the longitudinal, diametral and angular positioning.

Now modify the positionings in X and Z and the values for the angular orientation of the spindle in order to create the following part. It is very similar to the previous part but with variations on the width of the chamfers, the length of the turning, the longitudinal position of the holes, the radial depth of the holes and the angular orientation of the holes.

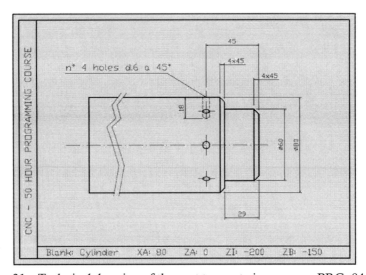

Fig. 31. Technical drawing of the part to create in program PRG_04_01

The arrow (→) before the block number in the program below shows you where to enter the appropriate value. Then check with the graphic simulation.

```
N10 ; blank part dimensions:
N20 ; XA = 80 bar diameter
N30 ; ZA = 0 machining allowance on front face
N40 ; ZI = -200 length of finished part
N50 ; ZB = -150 extension from jaws
N60
N70 WORKPIECE(,,,"CYLINDER",0,0,-200,-150,80)

N80 G18 G54 G90
N90 G0 X400 Z500
N100 M8
```

```
N110 SETMS(1)

N120 T1 D1 ; LEFT HAND ROUGHING TOOL
N130 G95 S1800 M4
N140
N150 ; COMPREHENSION OF THE X AND Z AXES LINEAR MOVEMENTS
N160
N170 ; DIAMETRAL MOVEMENT OF 20MM FROM X80 TO X60 MM
N180 G0 X80 Z0
N190 G0 X60
N200 ; THE REAL MOVEMENT OF THE TOOL IS 10 MM
N210
→ N220 ; LONGITUDINAL MOVEMENT OF ......... MM FROM Z0 TO Z-.........
→ N230 G1 Z-......... F0.2
→ N240 ; THE REAL MOVEMENT IS ......... MM
N250
N260 G0 X62 Z0
N270
→ N280 ; FOR THE CREATION OF A CHAMFER ......... x45
N290 ; THE MOVEMENT ON THE X-AXIS
N300 ; IS DOUBLE COMPARED TO THE MOVEMENT OF THE Z-AXIS
→ N310 G0 X......... ; INITIAL DIAMETER OF THE CHAMFER
→ N320 G1 X60 Z-......... ; FINAL VALUE IN Z
N330
→ N340 G1 Z-......... ; LENGTH OF THE TURNING
→ N350 ; IF THE SECOND CHAMFER IS ......... x45, THE INITIAL
→ N360 ; COORDINATE IN X IS......... MM BEFORE THE FINAL DIAMETER
→ N370 G1 X......... ; INITIAL DIAMETER
→ N380 G1 X80 Z-......... ; ARRIVAL POINT IN Z OF THE CHAMFER
N390
N400 G0 X200
N410 G0 Z200
N420
N430 ; COMPREHENSION OF THE VALUES FOR THE ANGULAR
ORIENTATION OF THE SPINDLE
N440
→ N450 ; CREATION OF 4 HOLES, STAGGERED BY ......... DEGREES
N460 ; POSITION OF THE FIRST HOLE AT ZERO DEGREES
N470 ; ANGULAR ORIENTATION OF THE SPINDLE AT 0 DEGREES
N480 SPOS=0
N490 SETMS(3)
N500
N510 T8 D1; RIGHT HAND RADIAL DRILL D.6
N520 G95 S1200 M3
→ N530 G0 Z-......... ; LONGITUDINAL POSITION OF THE HOLES
N540 STR_HOLE1:
```

```
N550 G0 X84
→ N560 G1 X……… F0.1 ; FINAL DIAMETER OF THE DRILL
N570 G4 S2
N580 G0 X84
N590 END_HOLE1:
N600
N610 ; ANGULAR POSITION OF THE SECOND HOLE
→ N620 SPOS[1]=………
N630 REPEAT STR_HOLE1 END_HOLE1
N640
N650 ; ANGULAR POSITION OF THE THIRD HOLE
→ N660 SPOS[1]=………
N670 REPEAT STR_HOLE1 END_HOLE1
N680
N690 ; ANGULAR POSITION OF THE FOURTH HOLE
→ N700 SPOS[1]=………
N710 REPEAT STR_HOLE1 END_HOLE1
N720
N730 G0 X200
N740 G0 Z200
N750
N760 M30
```

In order to correct the program wait until you have reached the end of this chapter.

4.10.3 Duplication, renaming and modification of a program

Normally, the programming of a new part does not start with an empty page. In order to reduce the time for starting the machine it is often decided to duplicate an existing program and modify it subsequently according to the new working cycle.

As the second program is very similar to an already existing program, it is duplicated and renamed. This procedure is very often used for numerically controlled machines.

Push PROGRAM MANAGER to access the program list, select the folder CHAP_04 with the arrows and open it with INPUT.

With the arrows, select the program to be duplicated: PRG_04_01. Press COPY. With the arrows, select the folder 01_EXERCISES where you want to insert the program and press PASTE.

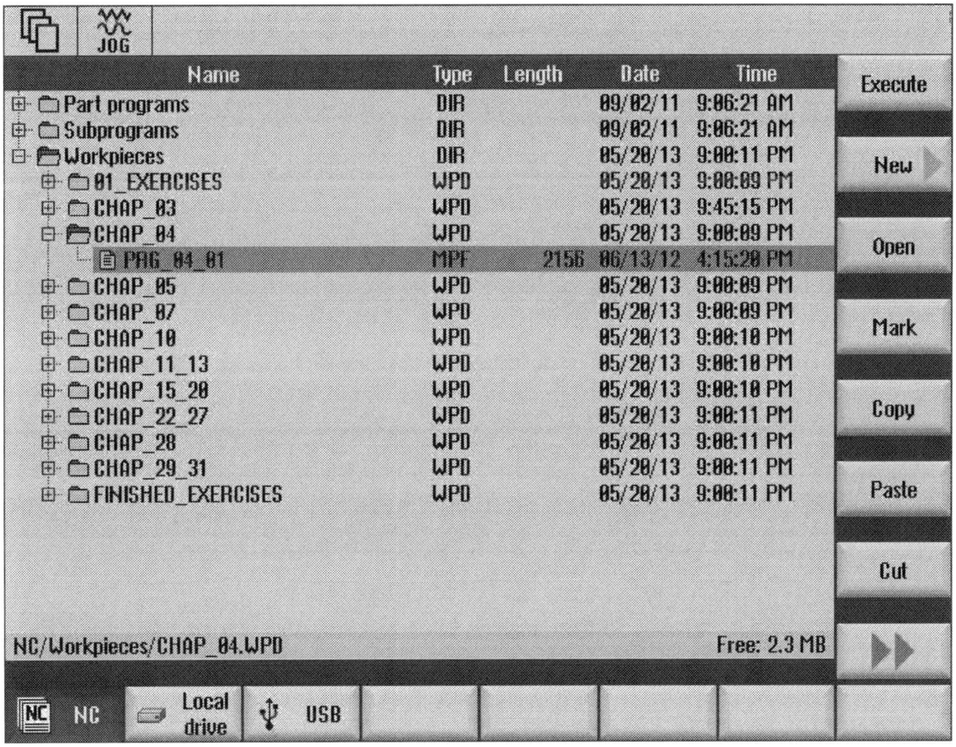

Fig. 32. Display of the PROGRAM MANAGER

The program is copied with the same name.
With the arrows, select the copied program, and press NEXT to rename it.

Then press PROPERTIES. A window with a summary of the file properties opens. With your mouse, click on the name of the program and change it into EX_04_01, then press OK to confirm.

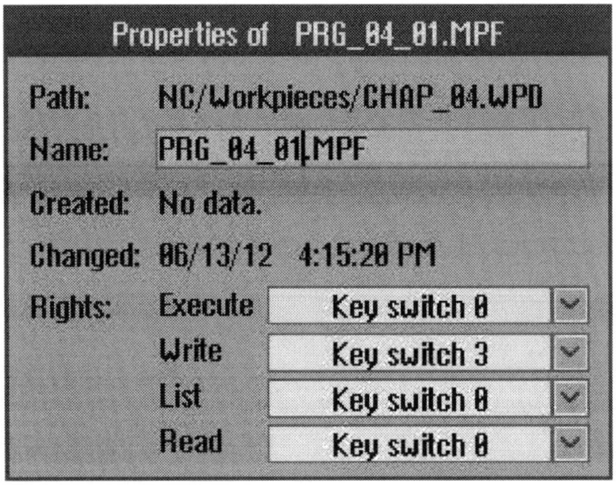

Fig. 33. File properties window

Now, open the copied file with INPUT and use the block numbers to transfer your conclusions from the book into the program, and update the comments.
Enable the single block mode and execute the graphic simulation.
Attention: before running the graphic simulation, be sure to set the execution speed of the profile to 100% by pushing 100% OVERRIDE as seen in paragraph 3.5.
Check the entered data and correct them if necessary. Use 3D VIEW, DETAILS and ROTATE VIEW to check the position of the holes.

Compare your program to the one in the folder FINISHED_EXERCISES named EX_04_01.

5. Programming Concepts (3h)
(Theory: 2h, Practice: 1h)

5.1 Elements constituting a program

The program consists of a sequence of instructions expressed by means of alphanumeric codes which give the machine all the necessary information in order to carry out a tooling operation.

The program is divided into a sequence of lines.

Every line is called "block" (e.g. G1 Z-20 G95 F0.1).

Every block contains one or more instructions defined as "words" (G1, Z-20, G95, F0.1).

Every word is made of a letter (G, Z, F) called "address", followed by a numeric value (1, -20, 95, 0.1).

The NC reads the program beginning with the first block, after the completion of the instructions therein it proceeds sequentially with the execution of the instructions entered in the following blocks until it arrives to the closing function of the program.

Block	Word	Word	Word	; Comment
Block	N10	G0	X20	; First block
Block	N20	G2	Z37	; Second block
Block	N30	G91	...	...
Block	N40	...	...	
Block	N50	M30	...	; End of program

Fig. 34. Name of the elements constituting the program

Every program has a name which is made of alphanumeric characters as well, there are no limits as far as its length is concerned, but only the first 24 characters can be displayed. The first two characters must be letters (or a letter with an underscore), followed by letters or numbers (e.g. _MPF100, SHAFT, SHAFT_2).

The sequence of the addresses programmed in a block does not impact the execution of the block itself. For a better understanding the following sequence is recommended:

$$N10 \ G... \ X... \ Y... \ Z... \ F... \ S... \ T... \ D... \ M...$$

The table below contains the first descriptions of some of the addresses which are most frequently used.

Address	Meaning
N	Address of block number
10	Block number
G	Preparatory function
X, Y, Z	Path information
F	Feedrate
S	Number of revolutions or cutting speed
T	Tool position
D	Number of tool corrector
M	Auxiliary function

Fig. 35. Meaning of some addresses

5.2 Logical programming sequence

When writing a program it is always recommendable to follow a precise logical sequence allowing not to forget any essential instruction.

The first element to be defined at the beginning of the program should be the **part zero point**, in other words the coordinate point X0 Z0 all the values saved in the program refer to; in a lathe, this point is often on the front face of the workpiece and on the rotating axis of the spindle.

Every single tooling operation is programmed as follows:
tool call, activation of the spindle, setting of the feedrate, rapid approach, execution of the tooling operation and disengagement from the workpiece with repositioning of the slide in the position where the tool is changed.

All the subsequent operations are programmed by repeating the same logical sequence.

5.3 Duration of the validity of an instruction

5.3.1 Modal instructions and groups of origin

Most instructions remain enabled in the blocks following the one where they were programmed; it is therefore not necessary to reprogram them until a function of the same type does not overwrite them. These functions are defined as modal functions.

The modal functions are deleted by functions belonging to the same group, i.e. functions which define similar instructions, but which contradict each other.

The function G1, which defines a working movement in linear interpolation, is deleted by the function G0, which belongs to the same group but defines a rapid movement.

The function G95, which sets the feedrate in millimeters per revolution, is deleted by the function G94, which belongs to the same group but sets the feedrate in millimeters per minute.

Below is the list of the most commonly used modal functions, subdivided by group of origin.

Name	Meaning
G0	Rapid traverse motion
G1	Linear interpolation
G2	Circular interpolation clockwise
G3	Circular interpolation counterclockwise
G33	Thread cutting with constant lead
G331	Rigid tapping
G332	Return (rigid tapping)
G34	Thread cutting with variable lead
G35	Thread with decreasing lead

Fig. 36. Group 1: Motion commands

Name	Meaning
G17	Plane selection 1st - 2nd geometry axis (X-Y)
G18	Plane selection 3rd - 1st geometry axis (Z-X)
G19	Plane selection 2nd - 3rd geometry axis (Y-Z)

Fig. 37. Group 6: Plane selection

Name	Meaning
G40	Deactivation of the tool radius compensation
G41	Activation of the tool radius compensation left of contour
G42	Activation of the tool radius compensation right of contour

Fig. 38. Group 7: Tool radius compensation

Name	Meaning
G500	Cancel all adjustable frames G54 - G57 if no value in G500
G54	Settable zero offset
G55	Settable zero offset
G56	Settable zero offset
G57	Settable zero offset

Fig. 39. Group 8: Settable zero offset (frame)

Name	Meaning
G60	Velocity reduction, precise stop
G64	Continuous path mode

Fig. 40. Group 10: Precise stop – continuous path mode

Name	Meaning
G70	Selects English units (inches and feet)
G71	Selects metric units (millimeter and meter)

Fig. 41. Group 13: Workpiece dimensioning inch/metric

Name	Meaning
G90	Absolute coordinate system
G91	Incremental coordinate system

Fig. 42. Group 14: Absolute/incremental coordinate system

Name	Meaning
G94	Linear feed mm/min or inch/min
G95	Rotational feed in mm/rev or inch/rev
G96	Constant cutting speed in m/min or feet/min
G97	Constant number of revolutions in rev./min

Fig. 43. Group 15: Feed rate and rotation type

5.3.2 Self-deleting instruction

Unlike the modal functions, **the self-deleting functions are valid only in the block where they are programmed**.
There are three self-deleting functions which are most widely used.
The first is G4, which sets the dwell time in seconds or revolutions; when the programmed dwell time ends, the function automatically disables and is not repeated in the next block.
The second is G9, which, entered into a certain block, sets an exact stop at the programmed arrival point.
The third is G53 and defines the machine coordinate system only in the block where it is programmed.

Name	Meaning
G4	Dwell time preset
G9	Exact stop only in the block where it is programmed
G53	Suppression of current frame

Fig. 44. Self-deleting instructions

5.4 Instruction types

The functions can also be grouped by type of command they set. Below are the most representative groups.

5.4.1 Technological instructions

Technological instructions are all those functions which define the cutting conditions.
Amongst these, we find the functions to select the position of the tool and define its corrector (s. fig. 35), those which set the cutting speed, the number of spindle revolutions and the feedrate of the tool (s. fig. 43).

5.4.2 Geometrical instructions

Geometrical instructions are linked to the definition of the reference system and to the tool path.
They define the type of tool path (fig. 36), the working planes (fig. 37), the activation of the automatic tool radius compensation (fig. 38), the reference system of the programmed values in the program (fig. 39), the exactness of the tool positioning (fig. 40 and 44), the type of measuring unit used (fig. 41) and the meaning of the numeric value (fig. 42).

5.4.3 Auxiliary instructions 'M'

The auxiliary instructions complete the information contained in a block.

With the 'M' functions it is for example possible to activate the cooling liquid, set the direction of the spindle rotation taking the back side of the spindle as reference, stop the program, set the end and other functions of the machine.

The meaning of most of the functions is decided by the manufacturer of the machine; it is therefore important to read the machine's manual in order to understand their meaning.

Below you will find a list of the instructions with fixed functionality used by most manufacturers.

Name	Meaning
M0	Programmed stop
M1	Optional stop activated by the control panel
M3	Spindle clockwise
M4	Spindle counterclockwise
M5	Spindle stop
M6	Tool change (if provided)
M8	Cooling liquid activation
M9	Cooling liquid stop
M30	End of program and return to beginning
M17	End of subroutine and return to main program
M40	Automatic gear change (when provided)
M41	Gear stage 1 (if provided)
M42	Gear stage 2 (if provided)
M43	Gear stage 3 (if provided)
M44	Gear stage 4 (if provided)
M45	Gear stage 5 (if provided)
M70	Spindle with transition to functioning as an axis

Fig. 45. Auxiliary or miscellaneous functions

5.5 Complementary instructions

In addition to the technological, geometrical and auxiliary functions there is a series of other commands which complete the programming. Below you will see how it is possible to enter comments, messages and the automatic numbering of the blocks into the program.

5.5.1 Entering of comments

In order to make the program clearer and more comprehensible, it is possible to enter comments.
Enter the comments at the end of the block and separate them from the block by means of a semicolon (;).
The comments are displayed together with the current block during the execution of the program.

```
N400 T1 D1 ;roughing tool
N410 X... Y...
N...
N500 T2 D1 ;finishing tool
N510 X... Y...
N...
```

5.5.2 Message display

The messages can be programmed to inform the operator of the tooling operation being currently executed.
The messages in the programs are created by writing the command MSG at the beginning of the block and then the message text between round brackets and quotation marks.

```
N400 MSG ("roughing") ;enabling of the message
N410 X... Y...
N...
N500 MSG () ;deletion of the message
```

The message text may have a maximum length of 124 characters, this will be displayed on two lines of 62 characters each.

5.6 Automatic block numbering

The block number is set by using the 'N' address, which identifies the position of the block in the program.

The number of the first block and the incrementation are defined by the programmer.

Before starting the graphic simulation it is always recommended to perform the automatic numbering of the blocks, which, in the event of a programming error, enables the NC to show the exact number of the line where the problem lies.

Normally, the number of the first block and the incremental value are both equal to ten; this allows to manually enter further blocks when modifying the program.

5.7 Practical exercise

5.7.1 Program analysis

Open the program 'PRG_05_01' in the folder 'CHAP_05' and look for all the elements of a program described in this chapter.
Start the graphic simulation in single block mode and analyze the movements of the tool according to the current program block with the help of the comments displayed.

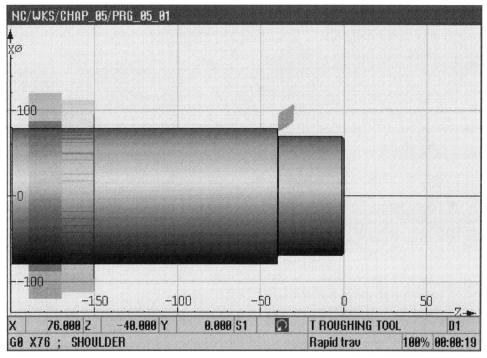

Fig. 46. Start of the simulation in order to analyze the program

```
; blank part dimensions:
; XA = 80 bar diameter
; ZA = 0 machining allowance on front face
; ZI = -200 length of finished part
; ZB = -150 extension from jaws

WORKPIECE(,,,"CYLINDER",0,0,-200,-150,80)

G18 G54 G90 ;G54 PART ZERO POINT SETTING
G0 X400 Z500
```

```
M8
SETMS(1)

T1 D1 ; LEFT HAND ROUGHING TOOL
G95 S1800 M4 ;SPINDLE ROTATION AND FEEDRATE SETTING
F0.2 ; FEEDRATE SETTING
G0 X82 Z0 ; APPROACH
G1 X-1; TOOLING OPERATION: FACING
G0 X66 Z2 ; DISENGAGEMENT
G0 Z0.5 ; APPROACH
G1 Z0 ; START OF TOOLING OPERATION
G1 X70 Z-2 ; EXTERNAL CHAMFER 2x45
G1 Z-40 ; TURNING
G0 X76 ; SHOULDER
G1 X82 Z-43 ; EXTERNAL CHAMFER 2x45
G0 X200 ; DISENGAGEMENT
G0 Z200
M30
```

5.7.2 Automatic block numbering

In this exercise you will learn the procedure for the automatic numbering of program blocks.
Copy the analyzed program (PRG_05_01) into the folder 01_EXERCISES following the steps described in chapter 4.10.3.
Change the name into EX_05_01.
Open it with INPUT.

Press NEXT:

Then RENUMBERING from the vertical softkeys:

The window for the block numbering appears:

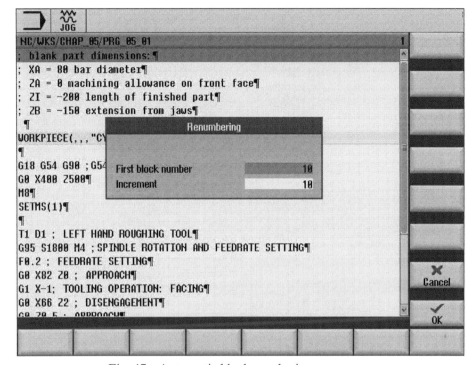

Fig. 47. Automatic block numbering screen

Enter the first number into the "First block number" field.
- Enter 10 and press INPUT.

In the "Increment" field, enter the incremental value for the numbering.
- Enter 10 and press INPUT.

- Press OK for the automatic numbering of the blocks.

Compare your program to the one in the folder FINISHED_EXERCISES named EX_05_01.

5.7.3 Deletion of block numbers

In order to delete block numbers go to the screen shown in fig. 47.
Enter '0' (zero) in both fields and confirm by pressing OK.
The numbering of the blocks will be deleted.

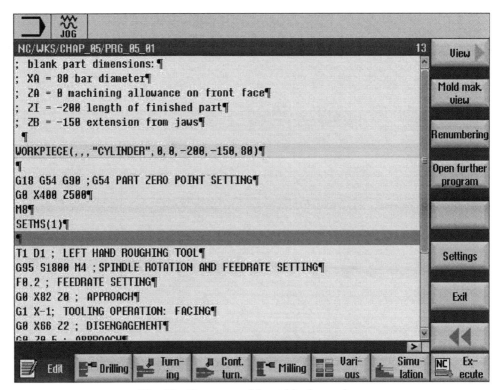

Fig. 48. Program without block numbers

6. Coordinate Systems (2h)
(Theory: 1h, Practice: 1h)

6.1 Machine coordinate system (MCS)

Every machine with a numeric control has a characteristic point to which the movements of the axes refer. This is called the **machine zero point**, i.e. the coordinate point X0, Z0, C0.

If there are no zero offsets enabled at the start of the machine, this is the only point the slides refer to.

All **slides have a characteristic point,** which is known to the NC; the coordinates displayed on the screen show the distance between the characteristic point of the slide and the machine zero point.

The reference system created is called machine coordinate system. This system is set by the manufacturer and can be modified by the operator through the program only.

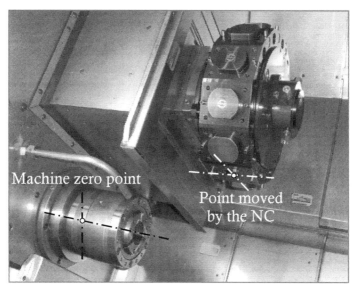

Fig. 49. Machine coordinate system: point moved by the NC referring to the machine zero point

The machine coordinate system is therefore independent of the length of the workpiece and the dimensions of the tool, it is not used during the processing of the workpiece but rather for the execution of safety positionings or generic positionings in the working area.

6.1.1 Machine zero point
The machine zero point of a lathe is usually to be found on the face of the spindle nose and on the main rotating axis (fig. 50.1).
The spindle nose is the centering element upon which are mounted the different applications for holding the workpiece.
The position of the machine zero point is chosen by the manufacturer who prefers a point firmly connected to the base of the machine, independently of the dimensions of the devices used to hold the workpiece (fig. 50.2.3.4).

Fig. 50. 1: Spindle nose ; 2: chuck with three jaws ; 3: Elastic collet for external hold; 4: Elastic expansion collet for internal hold of the workpiece

6.1.2 Characteristic point of the slide

The NC actually does not move the whole slide but only a characteristic point thereof. Its position compared to the machine zero point is the one displayed in the machine coordinate system.

The point moved by the NC is placed by the manufacturer in a position of the slide which is logical and easy to identify.

Undoubtedly, one of the most logical and easiest to find of turret positions is its center of rotation. Another point is the center of the attachment hole of the tool holder.

In the analyzed machine, as shown in fig. 49, the point moved by the NC lies at the center of the tool attachment hole (on the Z-axis) and on the external plane of the turret (on the X-axis).

The position of the point moved by the NC varies in different machines and it is very important for the operator to know where it is located; therefore always read the manufacturer's manual.

6.2 Workpiece coordinate system (WCS)

The machine coordinate system cannot be used for the definition of the tool path. **Every tooling operation is always programmed in the workpiece coordinate system, which you get by shifting the machine zero point onto the front face of the workpiece and the characteristic point of the slide onto the tool tip.**

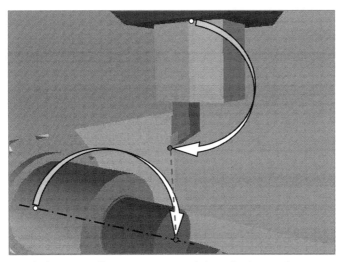

Fig. 51. Workpiece coordinate system: tool tip referring to the part zero point

6.2.1 G54 - G57: part zero point setting

The functions G54, G55, G56 and G57 offer the possibility to shift the machine zero point. The best position is on the front face of the workpiece. This new point is called the part zero point.

The part zero point is the coordinate point X0, Z0, to which refer all the values programmed after the enabling of one of the abovementioned functions.

In a lathe, the position X0 normally is not changed, but left on the rotating axis of the workpiece.

Instead, the zero point along the Z-axis is shifted, **the offset value on the Z-axis corresponds to the distance between the face of the workpiece and the machine zero point.**

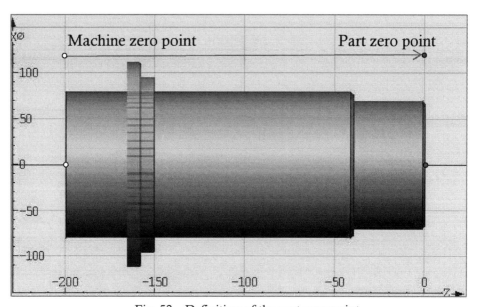

Fig. 52. Definition of the part zero point

The functions G54, G55, G56 and G57 are modal functions. They all belong to the same group and overwrite each other.

In every program, usually one (the most common being G54) function is used, or, if there are different clamping positions for the workpiece in the same cycle (first the front part is processed, the part is turned around and the back side is processed), two or more functions are programmed.

On the basis of the manufacturer's choices the function G54 may already be enabled at machine start and therefore it would not be necessary to enter it at the beginning of the program.

A table which is not part of the program contains the offset values which the operator will need to enter. **In order to get to this table press OFFSET on the control panel, then the horizontal softkey WORK OFFSET and then the vertical softkey G54... G57.**

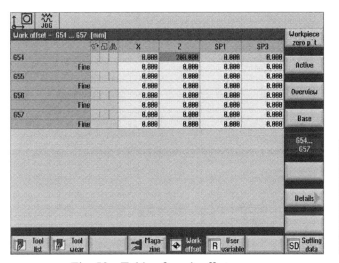

Fig. 53. Table of work offsets

In this table, next to the function, various columns can be found which offer the possibility to shift the machine zero point on all the axes set in the NC, including an angular value referring to the main spindle (SP1) and to the driven tools (SP3). In the second line, there's the possibility, for each function, to enter the correction to the main value.

It must be said that in most of the cases only the work offset on the Z-axis is used.

On the same page, by pressing the vertical softkey BASE, you will find the basic work offset, sometimes used by the operator to shift the machine zero point from the spindle nose to the front face of the jaws. The functions from G54 to G57 will later increment the 'basic' work offset up to the face of the workpiece.

The function G500, on the other hand, disables any work offset (when the manufacturer has not entered any value).

6.2.2 Tool offset

The tool path is described by programming the offsets of the tool tip in relation to the front face of the workpiece.

The tools used in the machine have different shapes and dimensions and this is something you have to tell your NC.

As we've already seen, the NC does not move a slide but rather a characteristic point thereof.

The position of the cutting edge is defined by the distance between the tool tip and the characteristic point of the slide on all the axes on which the slide moves (in this case X and Z).

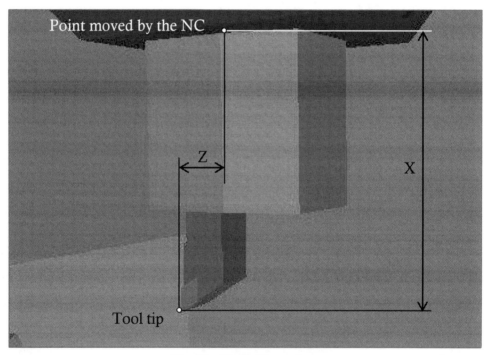

Fig. 54. Tool offset values

The offset values in X (for Siemens NCs) have a radial and not a diametral value, i.e. they correspond to the real distance between the tool tip and the point moved by the NC.

These values are entered by the operator in the table for the tool offsets where furthermore the insert radius and other data in relation to the graphic description of the tool, necessary for correct simulation, are specified.

6.3 Practical exercise

6.3.1 Setting of the part zero point, use of MDA and JOG

Before setting the part zero point it is necessary to know how to use the operating mode MDA, i.e. the operating environment allowing one to enter data manually.

The MDA is often used to perform small programs, to call tools into position and to enable functions like work offsets.

In order to measure the distance between the face of the workpiece and the machine zero point, select a tool offset, touch the face of the workpiece and copy the current position of the tool tip into the table of the work offsets, next to the function G54, in the Z column.

On the control panel, press MDA under the NC screen: the page showing the position of the axes and the cursor blinking and ready for the manual data entry appear.

Now enter T1D1, which is the function to call the tool into position and to activate the corresponding offset values.

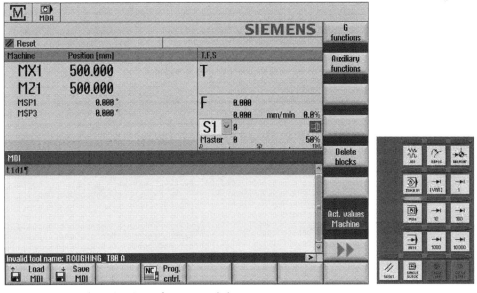

Fig. 55. Page for manual data entry

CNC – 50 Hour Programming Course

Press CYCLE START in order to execute the instructions. Now the turret has rotated to position the tool on the rotating axis of the spindle.

Press RESET ![RESET] to free the NC from the execution of the programmed block.

Now you are theoretically going to touch the face of the workpiece by moving the tool first in X and then in Z.

Push JOG.
Select the axis with which you want to move (X or Z).
Push the green button activating the spindle rotation.
Push the green button activating the feed.
With your mouse, bring the potentiometers (override) of the spindles and the feed to 100%.

Make sure that the LEDs show the activation of the buttons as shown in the following figure.

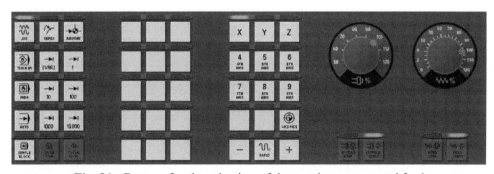

Fig. 56. Buttons for the selection of the continuous manual feed

Now move the selected axis with the plus and minus buttons.

Suppose you're going to touch the face of the workpiece now. If the axes move too quickly, reduce the feed potentiometer exactly as you would do on a real machine.

Bring the values of the axes to X30 and Z200 and make sure to display the values in the WCS as described in the following paragraph.

6.3.2 Display of the position in MCS and WCS

The current position of the slide can be displayed in machine coordinates or workpiece coordinates.

By pushing **ACT. VALUES MACHINE** from the vertical softkeys the reference system changes.

Disable the button and make sure to be in the workpiece coordinate system. Now take the slide to a theoretical position to touch the face of the workpiece (e.g. Z200). **In order to arrive exactly at Z200 select one of the buttons which set the feed by incrementation, expressed on the keys in thousandths of a millimeter (10, for example, means one hundredth at a time).**

Fig. 57. Buttons for the selection of the manual feed by incrementation

Enter this value (200) into the table for the work offsets, next to the function used in the program (G54) and in the Z column (see fig. 53).

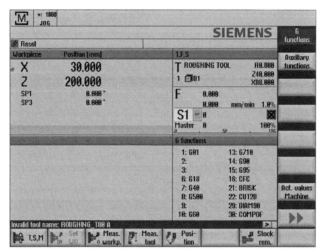

Fig. 58. Touch-off position of the part face in the workpiece coordinate system

Press MACHINE on the control panel to go back to the position of the axes. Press again ACT. VALUES MACHINE and select the machine coordinate system. The first difference you detect is that the position of X is expressed in a radial manner and not in a diametral manner as in the workpiece coordinate system. The second difference is that the position in X and Z varies exactly by the tool offset value contained in the geometry table (X=(30/2)+88=103).

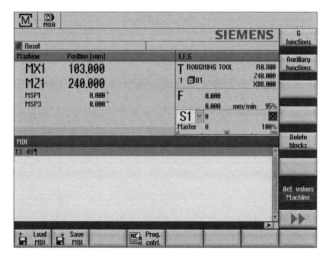

Fig. 59. Touch-off position of the part face in the machine coordinate system

Now return to MDA by pressing the relevant button.
Disable the button ACT. VALUES MACHINE.

Program G54 T1D1 and push CYCLE START (the functions for the newly set work offset and the tool offset values are being activated).

The current position of the tool on the Z-axis has become zero. As you can now see the part zero point is right on the front face of the touched workpiece.

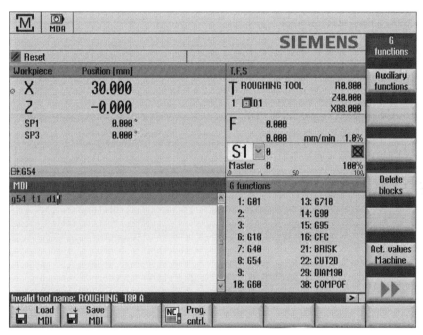

Fig. 60. Current position of the tool after activation of the workpiece coordinate system programmed in MDA

Attention: before going on press RESET to free the NC from **the execution of the programmed block.**

CNC – 50 Hour Programming Course

6.3.3 Tool offset by touching the workpiece

In order to get the tool offset values in X and Z it is possible to measure the distance between the tool tip and the point moved by the NC manually, as shown in fig. 54.

Another procedure often used consists in touching the workpiece, entering the value and letting the machine calculate the value automatically.

On the X-axis, the value to enter in order to do the calculation is the diameter where the tool has touched the workpiece.

On the Z-axis, the value to enter is the distance between the touched face and the activated work offset or the machine zero point.

Having a simulator at our disposal, the execution of an operating procedure would require an excessive effort of imagination which might confuse rather than teach.

It is sufficient to know that this procedure starts from the screen shown in fig. 61.

In order to access it, push OFFSET on the control panel, TOOL LIST select the tool to offset and press the softkey TOOL MEASURE.

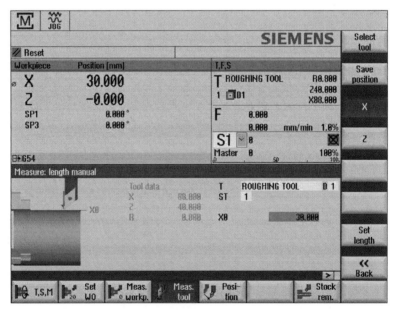

Fig. 61. Page for the automatic offset by touching the workpiece

7. Tool Call (2h)
(Theory: 1h, Practice: 1h)

7.1 Introduction
According to the logical programming sequence described in paragraph 5.2, after the setting of the part zero point, at the beginning of every operation, the tool and its corrector have to be selected.

7.2 T: tool call and function M6
By means of the address 'T', followed by the number of the position in which the tool is located, there is activation of the sequence of movements which permit the use of that tool for the tooling operation.

If the machine has a turret (as in this case), when reading the T instruction, the turret rotates to bring the tool onto the rotating axis of the workpiece.

It is useful to remember that the rotating movement of the turret is an operation with a high collision risk. For this reason, it is good practice, before calling the tool, to move the turret into a safety position, which is normally programmed in machine coordinates.

Other types of lathes have a tool magazine located outside the working area. This has a higher number of available positions, but needs more time for the tool change. In this case the procedure for the positioning of the tool does not consist in the simple rotation of the turret, but rather in a sequence of actions such as the automatic positioning at the coordinates for the tool change, the pneumatic release of the attachment cone, the movement of the magazine to the tool deposit position and the remaining procedure of selecting the programmed tool.

This long sequence of operations is usually activated by the auxiliary function M6, which in some machines needs to be combined with the tool call instruction.

The turret of the analyzed machine has 20 positions.
The most common lathes in industry have turrets with six, eight, or twelve positions.
The number of tools which can be mounted to the machine is very important, because it shows the maximum number of tooling operations that the lathe is able to perform in one working cycle.
To go to the tool list page, press OFFSET on the control panel and then TOOL LIST from the horizontal softkeys.

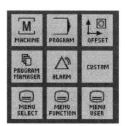

Fig. 62. Tool list page

The first column shows the physical location of the tool.
In the next practical exercise you will learn how to create and define, within the NC, the tools mounted to the turret.

7.3   D: tool offset values selection

For every tool it is necessary to define the table containing its offset values and the graphic description.
The address 'D' followed by the geometry table's progressive number activates the data contained therein.
These tables are also called CUTTING EDGES, because they set the position of the cutting edge controlled by the NC.

Various tables (up to 9) can be associated with every tool. This allows for the changing of the cutting edge used to describe the programmed profile.

The simplest example to explain the use of a tool combined with a double offset is that of a groove.

Often, in order to manage the width of the groove, users prefer first to command the left hand cutting edge and then the right hand cutting edge. The left hand cutting edge is defined in table D1 and then the right hand cutting edge in table D2, where only the geometry value in Z is modified by a value corresponding to the insert width.

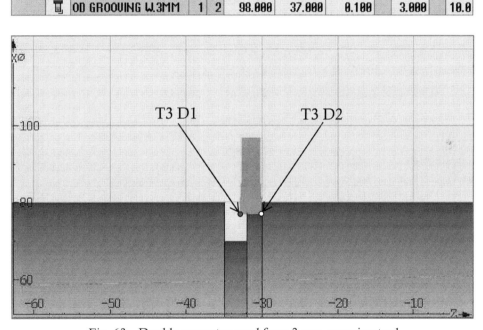

Fig. 63. Double corrector used for a 3 mm grooving tool

With the command D0 the tool offset values are deleted, with a return to command of the slide's characteristic point referring to the reference zero point active in that moment (part zero point or machine zero point). The following program (in the folder CHAP_07) creates a 5 mm groove, commanding first the left hand cutting edge (T3 D1) and then the right hand cutting edge (T3 D2).

```
; blank part dimensions:
; XA = 80 bar diameter
; ZA = 0 machining allowance on front face
; ZI = -200 length of finished part
; ZB = -150 extension from jaws

N10 WORKPIECE(,,,"CYLINDER",0,0,-200,-150,80)

N20 G18 G54 G90 ;G54 PART ZERO POINT SETTING
N30 G0 X400 Z500
N40 M8
N50 SETMS(1)

N60 T3 D1 ; GROOVING TOOL 3MM
; TABLE 1 DEFINES THE POSITION OF THE LEFT CUTTING
; EDGE
N70 G95 S1200 M4
N80 G0 Z-35
N90 G0 X82
N100 G1 X70 F0.12
N110 G0 X82

N120 D2 ; TABLE 2 DEFINES THE POSITION OF THE RIGHT CUTTING
; EDGE
N130 G0 Z-30
N140 G1 X70 F0.12
N150 G0 X82

N160 G0 X200
N170 G0 Z200
N180 M30
```

7.4 Correction of the tool wear

Every tool offset table is combined with a correction table, used by the operator to compensate the small variations on all axes in the machine due to the normal wear of the tool.

To go to the tool correction page press OFFSET on the control panel and then TOOL WEAR from the horizontal softkeys.

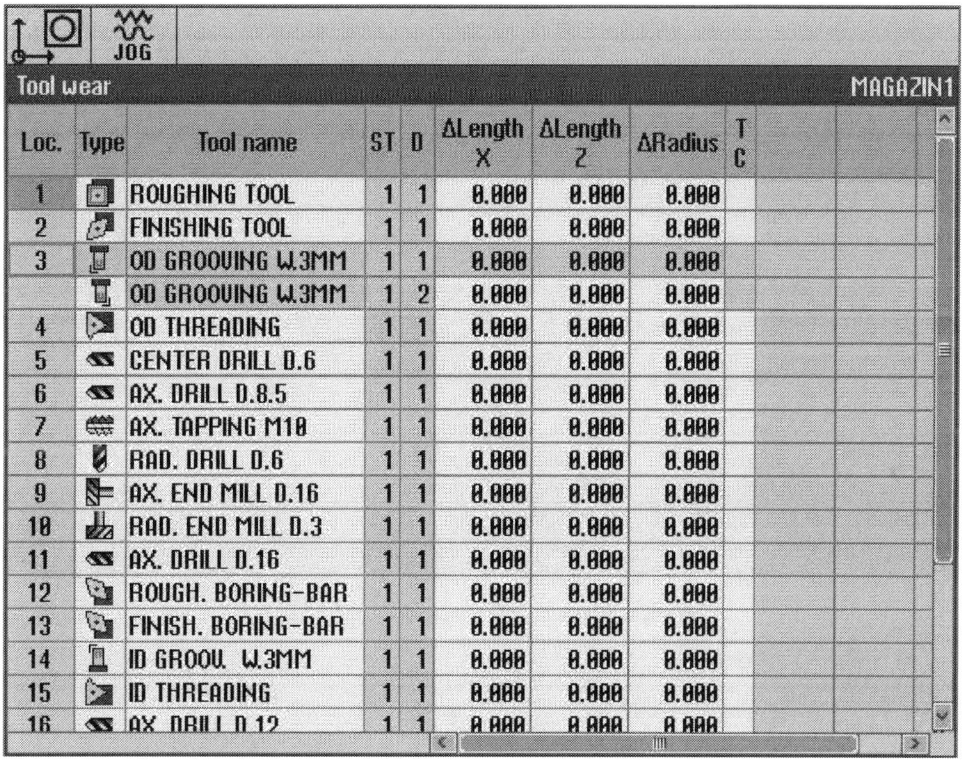

Loc.	Type	Tool name	ST	D	ΔLength X	ΔLength Z	ΔRadius	TC
1		ROUGHING TOOL	1	1	0.000	0.000	0.000	
2		FINISHING TOOL	1	1	0.000	0.000	0.000	
3		OD GROOVING W.3MM	1	1	0.000	0.000	0.000	
		OD GROOVING W.3MM	1	2	0.000	0.000	0.000	
4		OD THREADING	1	1	0.000	0.000	0.000	
5		CENTER DRILL D.6	1	1	0.000	0.000	0.000	
6		AX. DRILL D.8.5	1	1	0.000	0.000	0.000	
7		AX. TAPPING M10	1	1	0.000	0.000	0.000	
8		RAD. DRILL D.6	1	1	0.000	0.000	0.000	
9		AX. END MILL D.16	1	1	0.000	0.000	0.000	
10		RAD. END MILL D.3	1	1	0.000	0.000	0.000	
11		AX. DRILL D.16	1	1	0.000	0.000	0.000	
12		ROUGH. BORING-BAR	1	1	0.000	0.000	0.000	
13		FINISH. BORING-BAR	1	1	0.000	0.000	0.000	
14		ID GROOV. W.3MM	1	1	0.000	0.000	0.000	
15		ID THREADING	1	1	0.000	0.000	0.000	
16		AX. DRILL D.12	1	1	0.000	0.000	0.000	

Fig. 64. Tool correction page

It needs to be remembered that the potential use of a double corrector furthermore allows for correction of the tool when it is used for operations which require management independent of its wear, such as maintenance of a very tight tolerance during the finishing of multiple diameters.

7.5 Practical exercise

7.5.1 Creation of a tool

In all the exercises until this point we've used tools whose tooling data were imported from an external file according to the procedures laid down in paragraph 3.3.

It is now time to learn how to create new tools, how to set their offset values and the data used during the graphic simulation.

Press OFFSET, then press the TOOL LIST icon.
In order to display the icon which allows for the creation of a new tool it is important that no tool in the machine is selected.

Loc.	Type	Tool name	ST	D	Length X	Length Z	Radius		Loc. leng
1		ROUGHING TOOL	1	1	88.000	40.000	0.800 ←	93.0 55	11.0
2		FINISHING TOOL	1	1	94.000	40.000	0.200 ←	93.0 55	11.0
3		OD GROOVING W.3MM	1	1	98.000	40.000	0.100	3.000	10.0
		OD GROOVING W.3MM	1	2	98.000	37.000	0.100	3.000	10.0
4		OD THREADING	1	1	88.000	46.000	0.200		
5		CENTER DRILL D.6	1	1	100.000	24.000	6.000	118.0	
6		AX. DRILL D.8.5	1	1	100.000	56.000	8.500	118.0	
7		AX. TAPPING M10	1	1	100.000	81.000	10.000	1.500	
8		RAD. DRILL D.6	1	1	122.000	0.000	6.000	118.0	
9		AX. END MILL D.16	1	1	100.000	80.000	16.000	3	
10		RAD. END MILL D.3	1	1	134.000	0.000	3.000	2	
11		AX. DRILL D.16	1	1	100.000	120.000	16.000	118.0	
12		ROUGH. BORING-BAR	1	1	86.000	92.000	0.400 ←	93.0 55	8.0
13		FINISH. BORING-BAR	1	1	84.000	88.000	0.200 ←	93.0 55	8.0
14		ID GROOV. W.3MM	1	1	92.000	75.000	0.100	3.000	8.0
15		ID THREADING	1	1	88.000	95.000	0.200		
16		AX. DRILL D.12	1	1	100.000	72.000	12.000	118.0	

Fig. 65. Impossible to create a new tool when an already existing tool is selected

The icon with which to begin is NEW TOOL.

To make it appear, it is necessary to select an empty location in the turret or magazine.

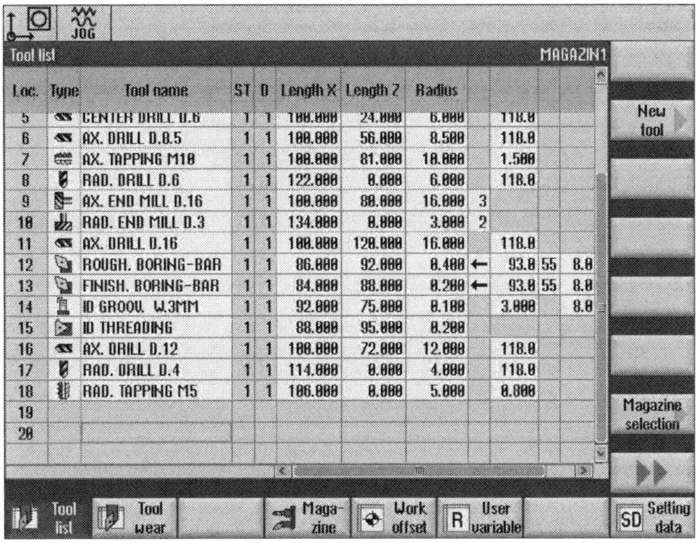

Fig. 66. Selection of an empty location for the creation of a new tool

Then press the vertical softkey NEW TOOL.

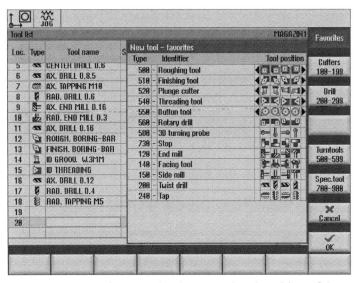

Fig. 67. Selection of the type of new tool to be created and position of the cutting edge

The NC now proposes different tool types, mainly divided by type of operation performed (in the favorite list you find turning, cutting and drilling tools and tools for special operations), graphic aspect and position of the cutting edge used during the operation.
The position of the cutting edge is extremely important for the graphic simulation.

Select a grooving tool with zero point on the lower left side (the drawing refers to the programming scheme specified in paragraph 4.9) and press OK.

For more options press the scrolling arrows:

Rename the newly created tool as 'EXAMPLE' in order not to exchange it or overwrite it with another existing tool.
Always confirm the changes by pressing INPUT.
Now set the offset value in X (e.g. 80 mm), then the offset value in Z (e.g. 40 mm), the value of the insert radius (e.g. 0.1 mm), the insert width (e.g. 3 mm) and its length (e.g. 10 mm).

Fig. 68. Creation of a new tool

Leave out the last three items in relation to the rotation direction of the spindle and the activation of the cooling liquid, as they are not used in ISO programming.

7.5.2 Deletion of a tool
In order to delete a tool, select the tool with the arrows, press the vertical softkey DELETE TOOL and confirm with OK.

7.5.3 Creation of a second tool corrector

As already seen in paragraph 7.3, a tool can be associated with various cutting edges. Now, define the secondary edge of a grooving tool according to the following procedure:

- Press OFFSET

- Make sure you are in the TOOL LIST

- With the arrows, highlight the tool you want to combine with another cutting edge

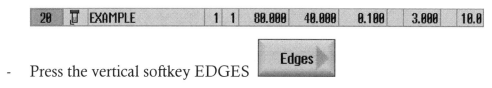

- Press the vertical softkey EDGES

- Press the vertical softkey NEW CUTTING EDGE

- A new window opens with the same tool offset values, though identified by the number 2 in column 'D'.

- Now select the box for the cutting edge, press SELECT and select the one down on the right.

- Now, vary the offset values according to the position of the new cutting edge. In this case, the second edge is located in the same position in X and changes in Z by a value which corresponds to the width of the insert. Enter the new value 37 (derived from: 40 − 3) in the Z column.

7.5.4 Deletion of a second tool corrector

In order to delete a second corrector formerly created:

- Select the edge you want to delete with the cursor.

20		EXAMPLE	1	1	80.000	40.000	0.100	3.000	10.0
		EXAMPLE	1	2	80.000	37.000	0.100	3.000	10.0

- Press the vertical softkey EDGES

- Then press the softkey DELETE CUTTING EDGE

Make sure not to delete a second corrector by pressing the softkey DELETE TOOL.

7.5.5 Mounting and removal of the tools in the turret

The lines from 1 to 20 represent the available locations on the turret. The following lines show all the tools created but not mounted to the turret, as if those were stored in an external magazine.

In order to remove a tool from the first 20 positions, select the tool you want to remove with the arrows and press the vertical softkey UNLOAD. In the opposite direction, to mount a tool archived in the following 20 locations, select the tool and press the vertical softkey LOAD. Automatically, a free location is proposed; change it if necessary and confirm with OK.

7.5.6 Saving of tooling data

Paragraph 3.3. contains the procedure to import tool data from an external file. Now we will see how to save these data in the folder containing the main program that has used them.
Press PROGRAM MANAGER on the control panel.
Select the folder in which you want to save the tooling data.
Press ARCHIVE in the list of vertical softkeys.
Then press SAVE SETUP DATA.

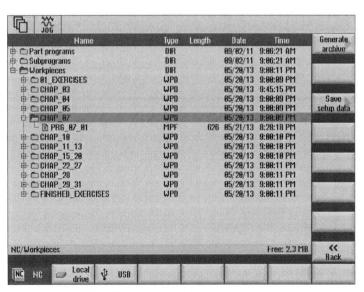

Fig. 69. Saving of tooling data

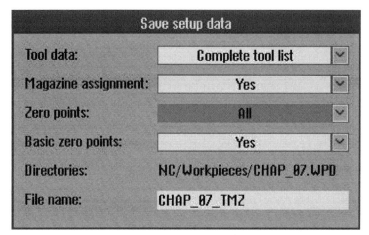

Fig. 70. Window for the saving of tooling data

Tool data: with the drop-down menu or the SELECT button placed at the center of the arrows, select *Complete tool list* to save the complete list of the tools in the machine. If you select *No* this means that you do not want to save the tool data but only the zero points (from G54 to G57) and the basic zero points (BASE).
Move with the arrows onto the next items.

Magazine locations: with the drop-down menu or the SELECT button, select *Yes*. This option saves the tools and their locations in the magazine. If you select *No*, the tool's location in the magazine is not saved.

Zero points: with the drop-down menu or the SELECT button, select *All*. This option saves the zero points (from G54 to G57). If you select *No* these data will not be saved.

Basic zero points: with the drop-down menu or the SELECT button, select *Yes*. This option allows for the loading of, not only the values for the zero points of the axes, but also the basic zero points.

Press OK to save the current data.

8. Spindle Activation (2h)
(Theory: 1.5h, Practice: 0.5h)

8.1 Introduction

The number of revolutions of the spindle and the diametral position of the tool are the data which define one of the most important cutting parameters of a tooling operation: the cutting speed.

The cutting speed is the speed with which the chip moves on the cutting edge of the tool.

The general concept of speed expresses the distance traveled in a certain unit of time.

In one revolution, the distance traveled by a point rotating on a certain diameter corresponds to the circumference that the diameter itself defines, calculated according to the formula shown in the following figure.

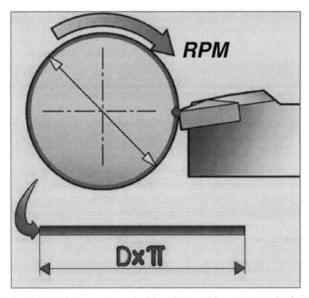

Fig. 71. Distance traveled by the tool in one revolution

By multiplying the circumference by the number of revolutions per minute of the spindle the total distance traveled by the tool in one minute is calculated.

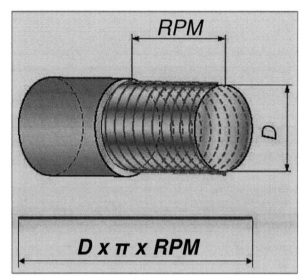

Fig. 72. Distance traveled by the tool in one minute with rotating workpiece

As the diameter is expressed in millimeters, this value is divided by one thousand so as to obtain the value for the cutting speed expressed in meters per minute, as in the formula shown in the following figure.

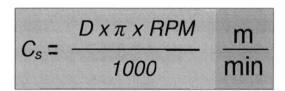

$$C_s = \frac{D \times \pi \times RPM}{1000} \quad \frac{m}{min}$$

Fig. 73. Formula for the calculation of the cutting speed

Learn these formula by heart as it is the basis of all turning technology.

8.2 SETMS: setting of the master spindle

In the program, before setting the rotation and feed in millimeters per revolution functions, it is necessary to define the spindle to which these refer.

The SETMS (Set Master Spindle) function, followed by the name of the spindle, defines the master or reference spindle.

On the page showing the current coordinates, other than the X- and Z-axis, the angular spindle positions are also shown: SP1 and SP3.

SP1 is the name of spindle '1' (workpiece carrying spindle).
SP3 is the name of spindle '3' (driven tools spindle).
SP2 would be the name of any counterspindle present in the machine.

Attention: Not all lathes use the same terms. Therefore refer to the manufacturer's manual.

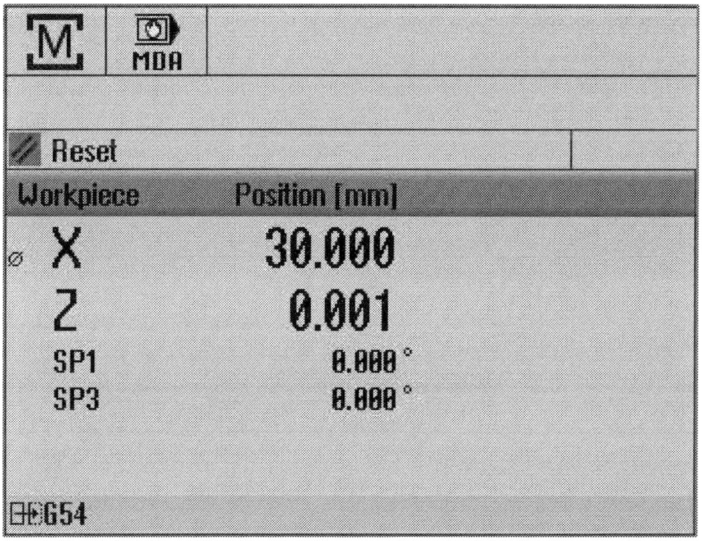

Fig. 74. Name of the spindles shown on the "current position" page

At the beginning of the programs that have been used so far, the SETMS(1) function is always programmed. This function sets the spindle holding the workpiece as master spindle, i.e. it says that all the rotation and feed functions (in mm/rev) programmed after its activation refer to spindle no. '1'.

8.3 G97: Spindle rotation with constant number of revolutions

The function G97 sets the spindle rotation to a constant number of revolutions. The number of revolutions is programmed using the address 'S'.

Example: **G97 S1000**.

The rotation direction of the spindle will be explained in paragraph 8.6.

The cutting speed depends on:
- material to be worked on (aluminium, steel, titanium, etc.),
- material of the tool used (HSS, sintered carbides)
- type of tooling operation (roughing, finishing, cutting, drilling)
- cutting conditions (workpiece extending much from spindle)

Its value, suggested by the tools' manufacturers or obtained from operator experience, is the only certain data item from which to start the calculation of the number of revolutions.

From the formula for the cutting speed, the inverse formula for the calculation of the number of revolutions to use on a certain diameter is obtained.

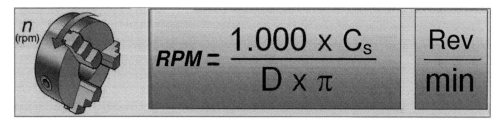

Fig. 75. Inverse formula for the calculation of the number of revolutions

We are now faced with the following question: What if, as happens with the facing or cutting of a workpiece, the diameter changes during the tooling operation?

The answer is that the cutting speed will change accordingly. With a constant number of revolutions, the cutting speed will be lesser on diameters smaller and greater on diameters larger than the one calculated, following the trend indicated by the graph in the following figure.

On the x-coordinate (the horizontal axis) we have the diameter value, on the y-coordinate (the vertical axis) we have the cutting speed value.

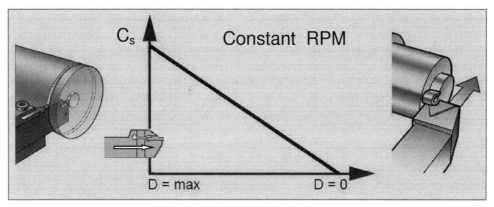

Fig. 76. Graph for the evolution of cutting speed at variation of working diameter, at constant number of revolutions

8.4 G96: setting of constant cutting speed

In order to maintain a constant cutting speed, a different function belonging to the same group as G97 needs to be used.

Indeed, calculation of the revolutions necessary to maintain constant cutting speed based on the diameter on which the tool is working is delegated to the machine by the function G96, followed by the address 'S' and by the value for the cutting speed (example: **G96 S120**).

After the programming of G96, the number of revolutions automatically adapts to the diameter of the workpiece: the higher the diameter is, the lower the number of revolutions is; the lower the diameter is, the higher the number of revolutions is.

If we now take the inverse formula for the calculation of the number of revolutions and if we perform a hypothetical facing operation up to a diameter of 2mm, we obtain as a result at constant cutting speed of 100 meters per minute a value for the spindle rotation of 15923 revolutions per minute.

If one were to consider the diameter to be zero, this value would even have to be infinite.

It is clear that the operator must be able to set the maximum number of revolutions achievable by the spindle.

This requirement leads us into the next paragraph.

8.5 LIMS=: limitation of the maximum number of revolutions

The command 'LIMS' sets a maximum limit for the number of revolutions of the master spindle (e.g. **LIMS=4000**).
Beyond this limit the cutting speed will inevitably be reduced as per the following graph.

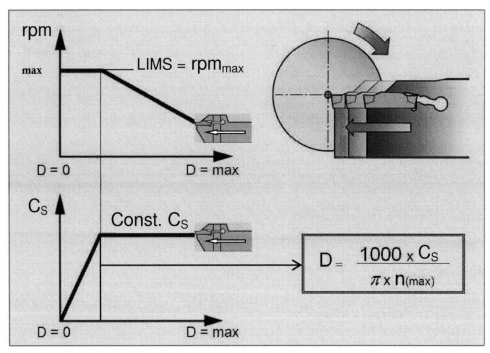

Fig. 77. Graph for the cutting speed trend beyond the number of revolutions threshold

Starting with the biggest diameter and approaching the smallest one, the NC holds the cutting speed constant and increases the number of revolutions up to the threshold set by the LIMS function.
From that point on, the cutting speed begins to drop starting from a diameter D (calculated by means of the formula shown in the figure) until it becomes zero ($C_s=0$) at the rotation center of the workpiece.

8.6 M3, M4, M5: setting of the rotation direction

The auxiliary functions M3 and M4 set the rotation direction of the spindle, in clockwise or counterclockwise direction.
Conventionally, the direction is defined from the point of view of an observer standing behind the spindle.
According to the structural features of the machine, the manufacturer suggests the ideal rotations direction.
What varies when turning in one direction rather than the other is the choice of tool type; the clockwise rotation (M3) requires the use of right hand drills, while the counterclockwise rotation (M4) requires the use of left hand drills.
The rotation direction is therefore also influenced by the characteristics of the tools available in the magazine.
The configuration of the machine under examination suggests a counterclockwise rotation (M4), because this allows the cutting stress to be discharged through the machine base and the insert to point towards the operator (simplifying the checking and replacement thereof).
The auxiliary function M5 stops the rotation of the spindle.

8.7 Instructions to a spindle which is not the master spindle

As already seen in paragraph 8.2 the SETMS function sets the master spindle.
This is the spindle to which refer instructions for speed ('S'), rotation direction ('M'), limitation of the number of revolutions ('LIMS=') and of angular orientation ('SPOS=').
If you want to use these functions in order to control a spindle which is not defined as the master spindle, or to display clearly the name of the spindle next to the function, you can use the following programming syntax:

```
G96 S1=120        ; S120 refers to the spindle named 1
G97 S1=2200       ; S2200 refers to the spindle named 1
G97 S3=1600      ; S1600 refers to the spindle named 3
M1=3              ; M3 refers to the spindle named 1
M1=4              ; M4 refers to the spindle named 1
M3=3              ; M3 refers to the spindle named 3
LIMS[1]=4000      ; LIMS=4000 refers to the spindle named 1
```

8.8 Choice of the functions G97, G96 and LIMS

The constant number of revolutions (G97) is programmed in the following cases:
- the diameter to be worked on does not change (cylindrical turning),
- fluctuations in the number of revolutions are not desired (execution of a thread in multiple passes),
- the working diameter is zero and therefore the number of revolutions would be incalculable (as for on-axis-drillings, where the number of revolutions is to be calculated according to the cutting speed and the diameter of the drill).

Constant speed is programmed (G96) when:
- the working diameter changes significantly (facing, parting-off, profiling of the workpiece).

The limit of the maximum number of revolutions is set (LIMS):
- every time that G96 is used in the program.

8.9 SPOS=: programming of the angular orientation

The option offered by a lathe to mount driven tools is always combined with its capability of orientating the spindle at an angle. This allows the execution of radial on-axis drilling and milling operations, or out-of-axis longitudinal (frontal) holes.

With SPOS it is possible to place the spindles in certain angular positions.

The simple angular orientation is not considered to be an axis on its own as it is not able to interpolate with any other axis in the machine (see paragraph 4.1).

The value that follows the SPOS function expresses the angle for the positioning of the spindle with reference to its zero point; this is expressed by a value between 0 and 360 degrees.

The programming is carried out as follows:

```
SPOS=0         ; angular orientation at zero degrees of the
               spindle defined as master spindle by means of the
               SETMS function

SPOS[1]=0      ; angular orientation at zero degrees of the
               spindle defined as no. '1' even if not defined as
               master spindle
```

8.10 Practical exercise

8.10.1 Calculation exercises
Based on the data referring to working diameter, number of revolutions and cutting speed, calculate and write down the missing data item in the relevant field.

Working diameter (mm)	Number of revolutions (r/min)	Cutting speed (m/min)
50	764	120
62	…………………	140
19	…………………	85
5	…………………	100
55	1200	…………………
8	1200	…………………
62	650	…………………
…………………	4500	100
…………………	2000	40
…………………	2000	220

Fig. 78. Exercise for the calculation of the cutting speed, the number of revolutions and the diameter from which the cutting speed begins to decrease

8.10.2 Creation of a new main program

The following paragraph is not bound to the topics covered in this chapter, as it describes the procedure for the creation of a new program.
Press PROGRAM MANAGER on the control panel.
With the arrows or with your mouse, select the system folder PART PROGRAMS.
This folder is designed only to contain main programs with .MPF (Main Program Files) extension.
Press NEW.
Press the vertical softkey PROGRAM GUIDE G CODE in order to create a program developed in ISO language, and not by means of the conversational software Siemens ShopTurn.
Enter the name of the new program (e.g. WORKPIECE_1).
Confirm with OK.
The newly created empty program opens automatically.

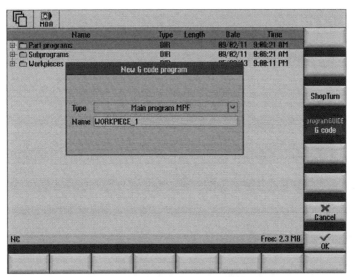

Fig. 79. Creation of a new program

Close the program by pressing NEXT 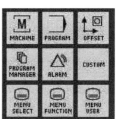 and then EXIT.
We are not going to use the PART PROGRAMS folder as it does not allow us to organize the programs into further subfolders.

8.10.3 Creation of a new subprogram

Press PROGRAM MANAGER on the control panel.
With the arrows or with your mouse, select the system folder SUBPROGRAMS.
This folder is designed only to contain subprograms called by main programs. Their extension is .SPF (Sub Program Files).
Press NEW.
Press the vertical softkey PROGRAM GUIDE G CODE in order to create a subprogram in ISO language.
Enter the name of the new subprogram (e.g. SUB_1).
Confirm with OK.
The newly created empty subprogram opens automatically.
Close it by following the procedure laid down in the previous paragraph.

8.10.4 Creation of a new folder

Press PROGRAM MANAGER on the control panel.
With the arrows or with your mouse, select the system folder WORKPIECES.
This folder is designed only to contain subfolders which can contain main programs and subprograms.
The extension of the folders is .WPD (Work Piece Directory).
Press NEW.
Enter the name of the new folder (e.g. FOLDER_1).
Confirm with OK.
The new folder will be listed together with the other folders in alphabetical order. A window will automatically open asking you if you want to create a main program with the same name in the folder.
You can choose to change the name, and, by means of the drop-down menu, select if you want it to be a main program (.MPF) or a subprogram (.SPF).
Maintain the selection MAIN PROGRAM MPF and press OK once again. Close the newly created program.
In order to create several .MPF or .SPF programs in the same folder, select the folder and proceed as described in paragraphs 8.10.2 and 8.10.3.

8.10.5 Spindle angular orientation exercises review

Repeat the exercises contained in paragraphs 4.10.1 and 4.10.2 and analyze the programming syntax for the angular orientation of the spindle.

9. Setting of the Feedrate (1h)
(Theory: 0.5h, Practice: 0.5h)

9.1 Introduction
The 'F' address expresses the value for the feedrate used during the tooling operations. Based on the instruction entered in the same block or on the active modal instruction, its value can be expressed in millimeters per revolution (G95) or in millimeters per minute (G94).

9.2 G95: feedrate expressed in mm/rev
The feedrate of a tool in a lathe is normally expressed in millimeters per revolution.
In this case the translation speed of the tool varies according to the number of revolutions of the spindle.

Attention: G95 also sets the fixed number of revolutions. This is why it often replaces function G97, as programmed in the following example block.

```
G95 S1800 M4      ; setting of the constant revolutions and
                  ; of the feedrate in millimeters per rev.
```

9.3 G94: speed in mm/min
G94 sets the feedrate of the tool as translation speed of the slide expressed in millimeters per minute.
This value remains totally unrelated to the number of revolutions of the spindle.
Multiplying the feedrate value expressed in millimeters per revolution (F_{rev}) by the number of revolutions per minute (RPM), the equivalent feed value expressed in millimeters per minute (F_v) is obtained.

$$F_v = F_{rev} * RPM$$

9.4 Calculation of the execution time for one pass

This paragraph analyzes the calculation method for the time the tool needs to execute one pass.

Assuming a cut lungth of 50 millimeters (L), a feedrate of 0.2 millimeters per revolution (F_r) and a spindle rotation of 1400 revolutions per minute (RPM), proceed as follows:

- Calculate the distance that the tool travels in one minute by using the formula shown in paragraph 9.3.

$$F_v = (F_r * RPM) = 0.2 * 1400 = 280 \text{ mm/min}$$

- From here you can calculate the speed expressed in millimeters per second.

$$v_s = F_v / 60 = 280 / 60 = 4.66 \text{ mm/sec}$$

- Dividing the length to be traveled by the execution speed of the pass, the time necessary to complete the pass is obtained.

$$t = L / v_s = 50 / 4.66 = 10.72 \text{ sec}$$

Combining these formulas you obtain:

$$t = L * 60 / (F_r * n)$$

This formula is useful in calculating in advance the time needed for the production of one piece.

9.5 Practical exercise

9.5.1 Calculation exercises

Based on the data referring to: length, feedrate and number of revolutions, calculate the time the tool needs to execute the pass.

Cut length (mm)	Feedrate (mm/rev)	Number of revolutions (r/min)	Time necessary (seconds)
60	0.3	840	....
60	0.12	1100	....
24	0.1	1260	....
18	0.06	780	....
22	0.14	1530	....
80	0.18	2100	....
66	0.05	1400	....
43	0.25	600	....

Fig. 80. Calculation exercises for the time needed for the tool to execute a pass

9.5.2 Saving of folders and programs

This paragraph explains how to save a folder or a program from the NC's memory on an external memory (usually with USB connection).

Before going on, connect a USB memory to the computer.

Press PROGRAM MANAGER on the control panel.

With the arrows or the mouse, select the program or the folder you want to save (e.g. the program PRG_03_01).

Press COPY from the list of vertical softkeys.
Then press USB in order to select the target device on which to save the data.
Press PASTE in order to start the copying process of the data.

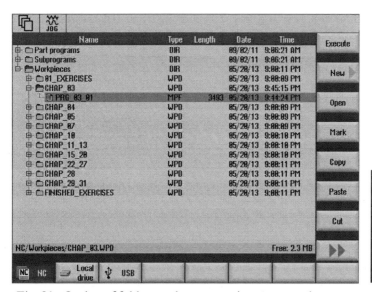
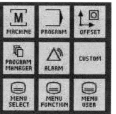

Fig. 81. Saving of folders and programs in an external memory

Press NC again in order to return to the NC's memory.

10. Absolute and Incremental Coordinates (1h)
(Theory: 0.5h, Practice: 0.5h)

10.1 G90: absolute programming
The function G90 sets the absolute coordinate system; this allows all coordinates expressed in the program to be referred to a single point, which can be the part zero point or the machine zero point. G90 is already enabled at the machine start up.

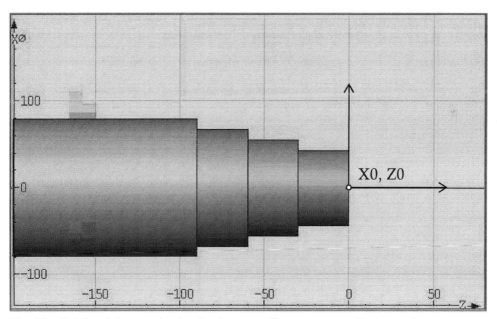

Fig. 82. Origin of the axes in the absolute coordinate system referring to the part zero

In order to support the operator in the programming of the profile, the designer often refers all the values to the front face of the workpiece.

The following drawing shows a workpiece with three steps of length 30 mm.

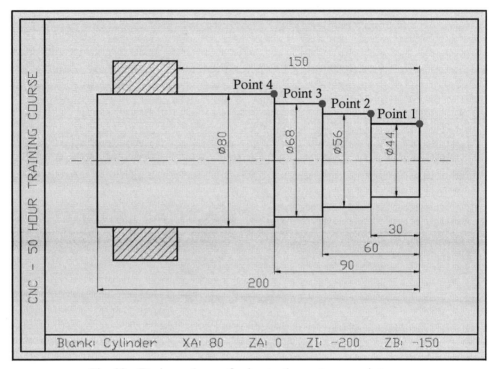

Fig. 83. Design values referring to the part zero point.

In the absolute coordinate system, in order to get from point 1 to the coordinate in Z of point 2, you program G1 Z-30, as the distance in Z of point 2 from the part zero point is 30 mm.

Now, in order to get from point 2 to the coordinate in Z of point 3, you program G1 Z-60, as the distance in Z of point 3 from the part zero point is 60 mm, although the distance between the two points is still 30 mm.

Now, in order to get from point 3 to the coordinate in Z of point 4, you program G1 Z-90, as the distance in Z of point 4 from the part zero point is 90 mm.

10.2 G91: incremental programming

The function G91 sets the incremental coordinate system.
When G91 is enabled, all coordinates refer to the current position of the tool.
The location of the tool becomes the zero point to which the subsequent movement refers.

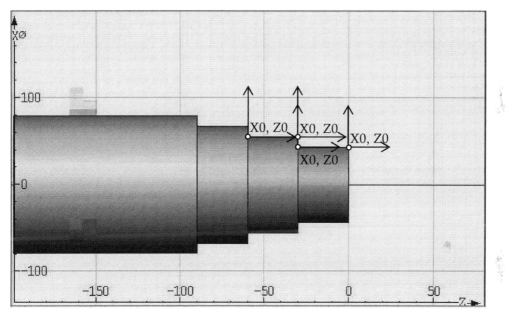

Fig. 84. Origin of the axes in the incremental coordinate system

The incremental coordinate system is never used for the definition of an entire profile. It is nonetheless useful (without ever being necessary) in defining the width of a groove or in programming a hole with chip removal.

Later we will see that incremental values are necessary when we want to repeat program parts, shifting their starting point (various holes executed with constant incrementation in Z), though in this case you can also use the mixed programming without necessarily enabling the function G91, as described in the following paragraph.

10.3 Mixed programming

By means of the function G90 it is possible to program incremental movements.

The possibility to enter values expressed both in absolute and in incremental coordinates in the same block is what gives this programming type its name.

The following programming syntax in most of the cases replaces the use of G91.

With function G90 enabled, to program an incremental movement of 60 mm on the Z-axis in its negative direction, you write:

$$G1\ Z=IC(-60)$$

In order to program an incremental movement of 5 mm on the X-axis in its positive direction, you write:

$$G1\ X=IC(5)$$

10.4 Diametral or radial meaning of the values associated with X

The enabling of the modal functions DIAMON, DIAM90 and DIAMOF determines diametral or radial meaning of the values programmed on the X-axis.

DIAMON attributes diametral meaning to all values associated with X.

DIAM90 sets diametral meaning in the absolute coordinate system (G90), while it attributes radial meaning when the value is expressed in the incremental coordinate system, both when it is enabled by function G91 or by means of the syntax X=IC(...).

DIAMOF attributes radial meaning to all values associated with X.

Attention: in the lathe that we are using, the active modal function at machine start is DIAM90.

10.5 Practical exercise

10.5.1 Analysis of a program in absolute coordinates

Open the program PRG_10_01 in the folder CHAP_10, start the graphic simulation and enable the single block mode. This program creates the part described in the drawing in fig. 83. Analyze the value of the programmed coordinates in every single block and monitor the tool's movement.

```
; blank part dimensions:
; XA = 80 bar diameter
; ZA = 0 machining allowance on front face
; ZI = -200 length of finished part
; ZB = -150 extension from jaws
N10 WORKPIECE(,,,"CYLINDER",192,0,-200,-150,80)

N20 G18 G54 G90 ;G54 PART ZERO POINT SETTING
; G90 ABSOLUTE COORDINATE SYSTEM
N30 G0 X400 Z500
N40 M8 ; COOLANT ACTIVATION
N50 SETMS(1) ; SETTING OF MASTER SPINDLE

N60 T1 D1 ; ROUGHING TOOL
N70 G95 S1800 M4 F0.2 ; SETTING OF NUMBER OF REVOLUTIONS AND
FEED RATE IN MM/REV

N80 G0 X68 Z2
N90 G1 Z-90
N100 G1 X82
N110 G0 Z2

N120 G0 X56
N130 G1 Z-60
N140 G1 X70
N150 G0 Z2

N160 G0 X44
N170 G1 Z-30
N180 G1 X58
N190 G0 Z2

N200 G0 X200
N210 G0 Z200
N220 M30
```

10.5.2 Analysis of a program in incremental coordinates

Open the program PRG_10_02 in the folder CHAP_10 and start the graphic simulation in single block mode. The same workpiece created on the previous page is now programmed in incremental coordinates using the mixed programming. Analyze the value of the programmed coordinates in every single block and monitor the tool's movement. Note how abuse of the incremental programming makes the program difficult to understand.

```
...
N20 G18 G54 G90 ;G54 PART ZERO POINT SETTING
; G90 ABSOLUTE COORDINATE SYSTEM
N30 G0 X400 Z500
N40 M8 ; COOLANT ACTIVATION
N50 SETMS(1) ; SETTING OF MASTER SPINDLE

N60 T1 D1 ; ROUGHING TOOL
N70 G95 S1800 M4 F0.2 ; SETTING OF NUMBER OF REVOLUTIONS AND
FEED RATE IN MM/REV

N80 G0 X68 Z2 ; POSITIONING IN ABSOLUTE COORDINATES
N90 DIAMON ; VALUE OF INCREMENTAL COORDINATES IN X WITH
DIAMETRAL MEANING

N100 G1 Z=IC(-92) ; ABSOLUTE COORD. Z-90
N110 G1 X=IC(14)  ; ABSOLUTE COORD. X82
N120 G0 Z=IC(92)  ; ABSOLUTE COORD. Z2

N130 G0 X=IC(-26) ; ABSOLUTE COORD. X56
N140 G1 Z=IC(-62) ; ABSOLUTE COORD. Z-60
N150 G1 X=IC(14)  ; ABSOLUTE COORD. X70
N160 G0 Z=IC(62)  ; ABSOLUTE COORD. Z2

N170 G0 X=IC(-26) ; ABSOLUTE COORD. X44
N180 G1 Z=IC(-32) ; ABSOLUTE COORD. Z-30
N190 G1 X=IC(14)  ; ABSOLUTE COORD. X58
N200 G0 Z=IC(32)  ; ABSOLUTE COORD. Z2

N210 G0 X200
N220 G0 Z200
N230 M30
```

11. Basic Functions to Define the Profile (3h)
(Theory: 1h, Practice: 2h)

11.1 G0: rapid movement

As already seen in the programs used so far, before the tooling operations, one or more blocks for approaching the tool to the workpiece are always programmed by means of function G0.

G0 sets the rapid movement of the slide or the tool to the programmed point.

The speed of the rapid movements is pre-set by the manufacturer and depends on the machine characteristics. For a lathe such as the one examined here, a maximum rapid movement speed of 30,000 mm/minute (30 meters per minute) is already ideal.

The modal function Siemens RTLION is active upon machine start and sets the linear trajectory of the rapid path.

The command RTLIOF overwrites this and sets the reaching of the arrival point without linear interpolation, thereby achieving a faster positioning speed but also increasing the risk of a collision.

When you enter this function into the program, be careful not to write GO (letter O) instead of G0 (number zero).

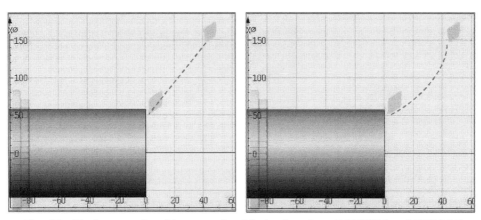

Fig. 85. Trajectory of rapid approach with the functions RTLION and RTLIOF

11.2 G1: linear interpolation

Function G1 sets a working movement to be executed in linear interpolation. The programmed point is reached describing a straight line starting from the point where the tool is located.

The feedrate used is the active modal feedrate or it is specified in the same block. If in the destination block one of the two coordinates does not change, it is not necessary to enter it again; in this case the movement will be only on the programmed axis.

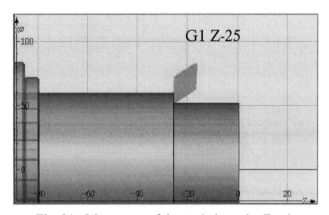

Fig. 86. Movement of the tool along the Z-axis

If, in the destination block, two values are programmed which differ from the starting point, the movement of the tool will occur along an inclined line, achieved by means of interpolation of the two axes (s. par. 4.8).

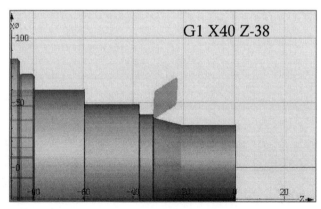

Fig. 87. Linear interpolation with tool moving along the axes X and Z

11.3 G33, G34, G35: threading in multiple passes

Function G33 sets a linear interpolation movement like the one set by G1, though synchronizing the start of the block with the angular zero position of the spindle.

This allows the execution of a threading with constant lead in multiple passes, during which the tool is always located in the same path, as shown in the next figure.

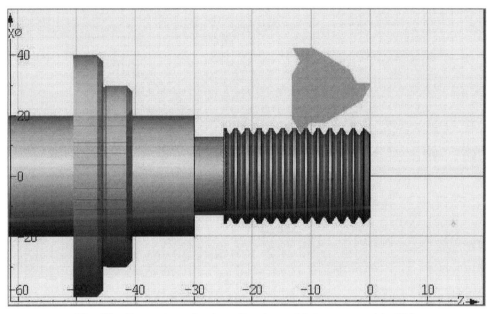

Fig. 88. Execution of a threading in multiple passes with G33

The lead of the thread is expressed in the same block as G33, using the address 'K' if the movement is on the Z-axis (G33 Z-20 K2), or 'I' (more unlikely) if the movement is on the X-axis.

The function G33 must always be programmed with a constant number of revolutions (G97/G95).

For conical threads, program the arrival point in X and Z and enter the value of the lead as a projection of the real lead on the predominant axis (Z for inclination angles under 45° and X for angles over 45°).

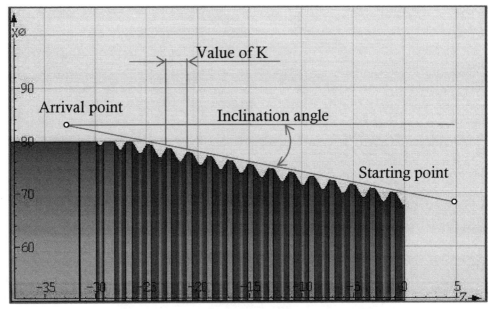

Fig. 89. Lead value to be programmed in a conical thread executed with G33

The programming of a thread with G33 is quite a long process. We will see below how to speed it up by using the automatic cycle CYCLE99.

For the creation of self-tapping screws with variable lead, program:
- G34: when the variation of the lead is progressively increasing, or
- G35: when the variation of the lead is progressively decreasing,

followed by the coordinates of the arrival point, the lead of the first spiral and the incremental value 'F' of the lead variation; for example

G34 Z-20 K2 F0.1

11.4 G4: dwell function

The function G4, when followed by the address 'F', sets a dwell time in seconds, or, if followed by the address 'S', in number of spindle revolutions. It is useful to guarantee the cylindricity of the bottom of grooves, in order to break or remove the chip during a drilling operation or in order to wait for a generic event (arrival of cooling liquid).

    G4 F1 ; dwell time of 1 second
    G4 S2 ; dwell time of 2 spindle revolutions

11.5 Practical exercise

11.5.1 Example of the roughing of a profile

Open the program PRG_11_01 in the folder CHAP_11_13, start the graphic simulation and enable the single block mode.
This program exclusively executes the roughing of the workpiece shown in the drawing below. Analyze the value of the programmed coordinates in every block and answer the questions in the paragraph below.

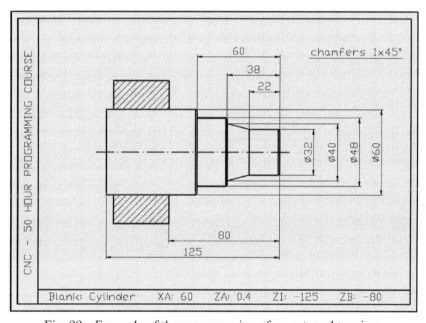

Fig. 90. Example of the programming of an external turning

```
; blank part dimensions:
; XA = 60 bar diameter
; ZA = 0.4 machining allowance on front face
; ZI = -125 length of finished part
; ZB = -80 extension from jaws
WORKPIECE(,,,"CYLINDER",0,0.4,-125,-80,60)

G18 G54 G90  ;G54 PART ZERO POINT SETTING
; G90 ABSOLUTE COORDINATE SYSTEM
G0 X400 Z500 ;SAFETY POSITION
M8 ; COOLANT ACTIVATION
SETMS(1) ; SETTING OF MASTER SPINDLE
```

```
T1 D1 ; ROUGHING TOOL SELECTION
G95 S1740 M4 ; SETTING OF NUMBER OF REVOLUTIONS AND FEEDRATE
IN MM/REV
G0 X52 Z2 ; RAPID APPROACH TO WORKPIECE
G1 Z-59.8 F0.2 ;FIRST PASS WITH FEEDRATE OF 0.2 MM/REV
G0 X54 Z2 ; RETURN OUTSIDE OF THE WORKPIECE FACE
G0 X46 ; RAPID POSITIONING AT DIAMETER 46
G1 Z-37.8 ;SECOND PASS
G1 X49
G1 Z-59.8
G0 X51 Z2
G0 X41 ;THIRD PASS
G1 Z-37.8
G0 X43 Z2
G0 X33 ;FOURTH PASS
G1 Z-21.8
G1 X41 Z-37.8
G0 Z0 ; POSITIONING FOR FACING
G0 X35 ; APPROACH TO DIAMETER
G1 X-1.6 ;EXECUTION OF FACING
G0 X60 Z2

G0 X200
G0 Z200
M30
```

Note how the final position of the facing is not X0 but X-1.6. Going beyond the center of the double insert radius avoids leaving the witness mark due to the presence of the radius itself.

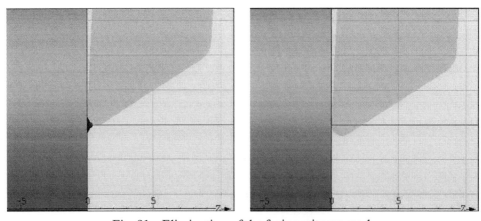

Fig. 91. Elimination of the facing witness mark

11.5.2 Review of program comprehension
Answer the following questions about the program you've just executed.

1) What is the cutting speed of tool T1 during the first pass?
   a) 120					b) 90					c) 100

2) What is the radial depth in X of the first pass?
   a) 5					b) 8					c) 4

3) At what value is diameter 48 roughed?
   a) 52					b) 49					c) 59.8

4) How much allowance is left on the shoulder at Z-38?
   a) 0.2					b) 0.1					c) 59.8

5) What are the coordinates of the position where the tool is located before executing the facing?
   a) X35, Z0				b) X41, Z-0.4				c) X62, Z0

The correct solutions can be found in the program ANS_11_01 in the folder CHAP_11_13.

11.5.3 Example of the programming of a threading

Open the program PRG_11_02 in the folder CHAP_11_13, start the graphic simulation and enable the single block mode.
This program executes:
- with tool T1 D1, the facing, the chamfer and the external turning of the workpiece,
- with tool T3 D1 (with 3 millimeter insert) the groove at the end of the thread,
- with tool T4 D1, the thread in multiple passes.

The number of passes in order to execute a thread depends on the dimensions of its lead. The depth of every pass is recommended by the tools' manufacturer according to the material to be worked on.
In this case, 8 passes have been executed, the depth of which is shown in the program.

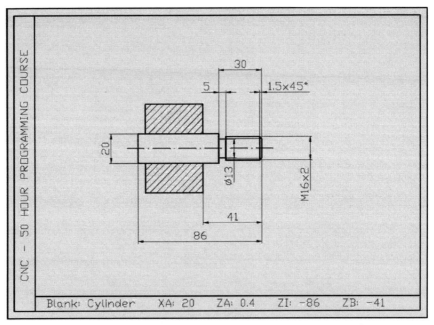

Fig. 92. Example of the programming of a threaded workpiece

```
; blank part dimensions:
; XA = 20 bar diameter
; ZA = 0.4 machining allowance on front face
; ZI = -86 length of finished part
; ZB = -41 extension from jaws
```

```
WORKPIECE(,,,"CYLINDER",0,0.4,-86,-41,20)

G18 G54 G90 ;G54 PART ZERO POINT SETTING
; G90 ABSOLUTE COORDINATE SYSTEM
G0 X400 Z500 ;SAFETY POSITION
M8 ; COOLANT ACTIVATION
SETMS(1) ; SETTING OF MASTER SPINDLE

LIMS=3000 ; MAXIMUM LIMIT OF REVOLUTIONS
T1 D1 ; EXTERNAL TURNING
G96 S100 M4 ; SETTING OF CONSTANT CUTTING SPEED AND FEEDRATE
IN MM/REV
G0 X22 Z0 ; RAPID APPROACH TO WORKPIECE
G1 X-1.6 F0.18 ; FACING
G0 X12.8 Z0.5 ; DIAMETER AT THE BEGINNING OF THE CHAMFER
G1 Z0 ; APPROACH TO FACE OF WORKPIECE
G1 X15.8 Z-1.5 ; EXECUTION OF THE CHAMFER 1.5 X 45
G1 Z-30 ; TURNING
G1 X22 ; STRAIGHT SHOULDER
G0 X200 ; DISENGAGEMENT IN X
G0 Z200 ; DISENGAGEMENT IN Z

T3 D1 ; GROOVING TOOL WIDTH 3MM
G95 S800 M4 ; SETTING OF NUMBER OF REVOLUTIONS AND FEEDRATE
IN MM/REV
G0 Z-30 ; RAPID POSITIONING IN Z
G0 X22 ; APPROACH TO DIAMETER OF BAR
G1 X13 F0.1 ; EXECUTION OF THE GROOVE
G4 S2 ; DWELL TIME OF TWO REVOLUTIONS AT BOTTOM OF GROOVE
G0 X22
G0 Z=IC(2) ; INCREMENTAL MOVEMENT OF 2MM IN Z POSITIVE
G1 X13
G4 S2
G0 X22
G0 X200 ; DISENGAGEMENT IN X
G0 Z200 ; DISENGAGEMENT IN Z

T4 D1 ; TOOL FOR EXTERNAL THREADS
G95 S600 M3 ; INVERSION OF SPINDLE ROTATION DIRECTION
G0 Z4 ; RAPID POSITIONING IN Z

; POSITIONING AT DIAMETER OF FIRST PASS BEGINNING FROM A
NOMINAL DIAMETER OF 16 MM
G0 X15.4 ; DEPTH OF RADIAL PASS OF 0.3MM
G33 Z-29.5 K2
G0 X18 ; EXIT FROM THREAD
G0 Z4 ; REPOSITIONING IN Z
```

```
G0 X14.9 ; DEPTH OF RADIAL PASS OF 0.25MM
G33 Z-29.5 K2 ; SECOND PASS
G0 X18 ; EXIT FROM THREAD
G0 Z4 ; REPOSITIONING IN Z

G0 X14.5 ; DEPTH OF RADIAL PASS OF 0.2MM
G33 Z-29.5 K2 ; THIRD PASS
G0 X18 ; EXIT FROM THREAD
G0 Z4 ; REPOSITIONING IN Z

G0 X14.1 ; DEPTH OF RADIAL PASS OF 0.2MM
G33 Z-29.5 K2 ; FOURTH PASS
G0 X18 ; EXIT FROM THREAD
G0 Z4 ; REPOSITIONING IN Z

G0 X13.8 ; DEPTH OF RADIAL PASS OF 0.15MM
G33 Z-29.5 K2 ; FIFTH PASS
G0 X18 ; EXIT FROM THREAD
G0 Z4 ; REPOSITIONING IN Z

G0 X13.56 ; DEPTH OF RADIAL PASS OF 0.12MM
G33 Z-29.5 K2 ; SIXTH PASS
G0 X18 ; EXIT FROM THREAD
G0 Z4 ; REPOSITIONING IN Z

G0 X13.36 ; DEPTH OF RADIAL PASS OF 0.10MM
G33 Z-29.5 K2 ; SEVENTH PASS
G0 X18 ; EXIT FROM THREAD
G0 Z4 ; REPOSITIONING IN Z

G0 X13.26 ; DEPTH OF RADIAL PASS OF 0.05MM
G33 Z-29.5 K2 ; EIGHTH PASS
G0 X18 ; EXIT FROM THREAD

G0 X200 ; DISENGAGEMENT
G0 Z200
M30
```

11.5.4 Finishing of a profile
This exercise allows for the consolidation of the learning of much of the information delivered up to now.
- Open the program PRG_11_01 in the folder CHAP_11_13
- copy it into the folder 01_EXERCISES according to the procedures laid down in paragraph 4.10.3
- change the name into EX_11_01
- **at the end of the program, after the roughing, enter the finishing of the profile according to the programming sequence specified in paragraph 5.2**
- the part to be created is described in the drawing in fig. 90
- make sure the starting values of the chamfers in X (paragraph 4.10.1) are correctly programmed

Compare your program to the one in the folder FINISHED_EXERCISES named EX_11_01.

12. Direct Programming of Rounds, Chamfers and Angles (2h)
(Theory: 1h, Practice: 1h)

12.1 Introduction
By now, we've seen that all segments constituting a profile are defined by programming the coordinates of their arrival point.
There is also a simplified programming method, which delegates to the NC the calculation of the tool path by directly programming rounds, chamfers and inclination angles of the lines compared to the main rotating axis.

12.2 RND= / RNDM=: execution of a round
The function RND, followed by the value of the radius, allows for the entry of a tangential round between linear and circular parts of the profile at the end of a block.

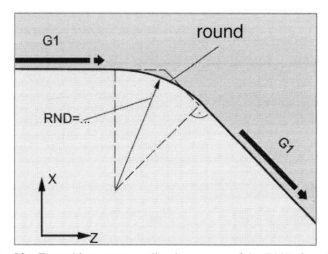

Fig. 93. Round between two line by means of the RND function

The starting and arrival point of the round depend on the dimensions of the programmed radius and the direction of the two blocks to be connected.

RND is not a function designed for the programming of a circle arc; it is instead designed to simplify the programming of the breaking of a sharp edge by a round radius.

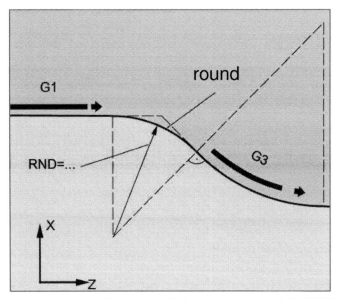

Fig. 94. Round between a line and a circle arc by means of the RND function

The RND function is self-deleting, therefore it is executed only in the block where it is programmed.

In order to combine multiple consecutive edges, use the modal function RNDM which is deleted by RNDM=0.

Programming syntax:

```
G1 Z-20 RND=0.4
G1 X18 Z-14
```

or:

```
G1 Z-20 RNDM=0.4
G1 X18 Z-14
```

12.3 CHR= / CHF=: execution of a chamfer

As for the function RND, the function CHR, followed by the dimensions of the chamfer, allows for the entry of the edge breaking located at the intersection between linear or circular parts of the profile at the end of a block.

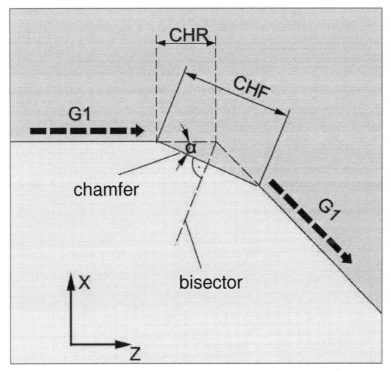

Fig. 95. Chamfer executed between two lines by means of the function CHR or CHF

The dimension of the chamfer is defined with CHR when the value is expressed according to the direction of the segments, or with CHF when the value refers to the real length of the chamfer.

The starting and the arrival point of the chamfer depend on the type of programmed function (CHR or CHF), on the dimensions of the chamfer and on the direction of the two profile blocks.

CHR and CHF are not functions designed for the programming of an inclination angle of the chamfer; they are instead designed to simplify the programming of the breaking of a sharp edge.

12.4 FRC= / FRCM: specific feedrate on chamfers and rounds

In order to optimize the surface quality of chamfers and rounds it is possible to program by means of the function FRC (Feedrate on Round and Chamfer) a specific feedrate with which to create them.

FRC, followed by the feedrate value, needs to be entered in the same block in which you program radii or rounds.

FRC is a self-deleting function, therefore it is executed only in the block where it is programmed.

The value following the function needs to be expressed in the measuring unit in relation to the enabled feedrate function (G95 or G94).

In order to program a specific modal feedrate, i.e. enabled for all rounds and/or chamfers programmed subsequently, use the function FRCM; to disable program FRCM=0.

Programming syntax:

```
G1 Z-20 CHR=1 FRC=0.02
G1 X18 Z-14
```

or:

```
G1 Z-20 RND=4 FRC=0.02
G1 X18 Z-14
```

or:

```
G1 Z-20 RND=4 FRCM=0.02
G1 X18 Z-14
```

(all the subsequent rounds and chamfers, programmed with RND, CHR and CHF, are executed with the specific feedrate of 0.02 millimeters per revolution)

12.5 ANG=: direction of a line defined by an angle

The function ANG allows for the definition of profile lines by directly programming the inclination angle value of the line with respect to the positive direction of the Z-axis.

The value of the angle to be programmed is obtained by using the scheme below, already presented in paragraph 4.9.

Position the tool tip at the center of the Cartesian axes; the direction of the path you want to execute indicates the inclination angle to be programmed.

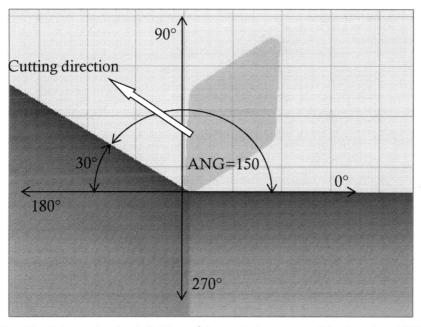

Fig. 96. Scheme for the definition of the angle by means of the function ANG

The block describing the line must contain one single coordinate of the arrival point (either X or Z) and the inclination angle with which to reach it.

Programming syntax: G1 Z-40 ANG=150

It is furthermore possible to program in a block only the cutting direction and in the next block the coordinates of the arrival point in X and Z together with the angle value.

Programming syntax:
```
G1 ANG=180
G1 Z-38 X40 ANG=166
```

The arrival point of the first block is calculated by the NC on the basis of the position of the second point and the direction of the two lines.

At the end of the block it is possible to program radii or chamfers with the functions RND, RNDM, CHR, CHF.

12.6 Practical exercise

12.6.1 Point to point and direct programming comparison

In paragraph 11.5.4 we executed the finishing program of the profile by always entering the coordinates of the arrival point.
Now, the same drawing will be used, the cone though is defined by means of only the arrival point in Z together with its inclination angle.

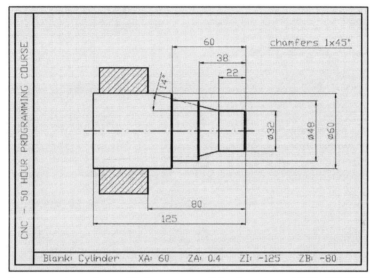

Fig. 97. Programming of a profile by means of the functions CHR, FRCM and ANG

Open the program PRG_12_01 in the folder CHAP_11_13.
In this program no changes have been made to the roughing path, but the finishing is programmed using the direct programming functions for chamfers and angles.
Start the graphic simulation and enable the single block mode: analyze the programmed functions and the respective tool movement.

Compare the new program on the next page with the previous one.
You will see that for the execution of the frontal chamfer it is necessary to change the starting point and program a vertical line which intersects with the next block at an angle of 90°.

The value of the inclination angle to be programmed for the execution of the cone refers to the positive direction of the Z-axis as shown in figure 96 and in the programming scheme in paragraph 4.9.

Previous program created by programming the arrival point coordinates.	New program created by the direct programming of chamfers and angles.
;FINISHING OF THE PROFILE T2 D1 G95 S1800 M4 **G0 X30 Z2** G1 Z0 F0.1 G1 X32 Z-1 G1 Z-22 G1 X40 Z-38 G1 X46 G1 X48 Z-39 G1 Z-60 G1 X58 G1 X60 Z-61 G1 Z-62 G1 X61 G0 X200 G0 Z200 M30	;FINISHING OF THE PROFILE T2 D1 G95 S1800 M4 **G0 X26 Z2** G1 Z0 F0.1 G1 X32 **CHR=1 FRCM=0.04** G1 Z-22 G1 Z-38 **ANG=166** G1 X48 **CHR=1** G1 Z-60 G1 X60 **CHR=1** G1 Z-62 G1 X61 G0 X200 G0 Z200 M30

Fig. 98. Comparison between the two programs which create the same profile: in the left column by means of the point to point programming, in the right column by using the direct programming functions CHR, FRCM and ANG

In order to use the direct programming functions for chamfers and rounds it is recommended that the block where they are programmed be followed by a working movement and not by a rapid movement.

The space defined between the starting block and the arrival block must be sufficient to contain the dimensions of the programmed chamfer or round.

12.6.2 Definition of the blank part data

The dimensions of the blank part are used by the graphic simulation in order to display the part to be worked on.
The blank part data need to be entered at the beginning of the program.

Before that, it is recommended to write down comments which report their dimensions, as has been done in all the programs used until now.

Following the procedure specified in paragraph 8.10.2, create a new main program (.MPF) in the folder 01_EXERCISES and name it EX_12_01.
At the beginning, the program is empty. Enter the comments which describe the dimensions of the blank part with reference to the drawing in figure 100.

```
; blank part dimensions:
; XA = 50 bar diameter
; ZA = 0.3 machining allowance on front face
; ZI = -100 length of finished part
; ZB = -70 extension from jaws
```

The table of blank data shown by the NC identifies the basic parameters for its definition with the letters XA, ZA, ZI and ZB. Their meaning is specified in paragraph 3.4.

Place the cursor on the line after the comments in order to enter the blank data.

Press the horizontal softkey VARIOUS.

Then press the vertical softkey BLANK.

Enter the values and confirm with ACCEPT.

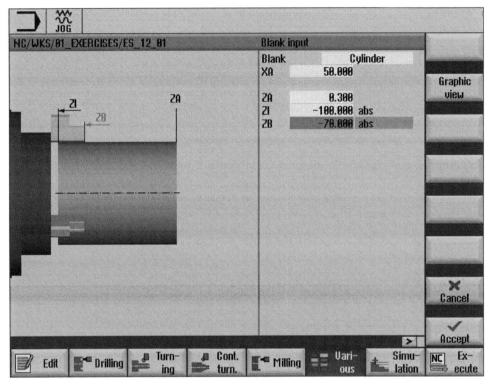

Fig. 99. Page for entering the blank part data

In order to return and modify the values after their acceptance, press the arrow shown at the end of the block with your mouse.

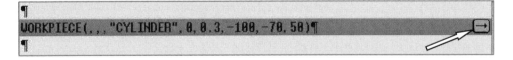

Attention: with the SELECT button it is possible to set the value of ZI and ZB with absolute or incremental meaning. The selection of 'absolute' shows that the value expressed refers to the part zero point; the selection of 'incremental' shows that the value expressed refers to the front face of the workpiece inclusive of machining allowance set at parameter ZA.

12.6.3 Programming of a workpiece

Enter the missing data into the program below.
The arrow (→) before the block number tells you where to enter the value.
After entering the values, write the complete text in the newly created program EX_12_01.

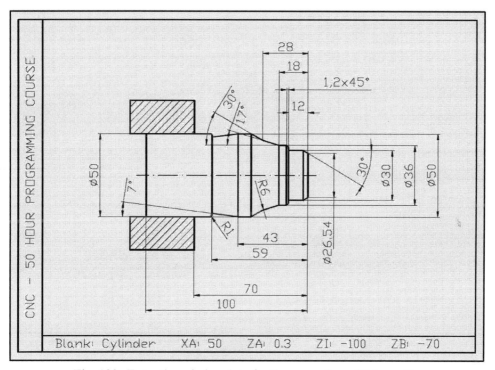

Fig. 100. Enter the missing data for the execution of this profile

```
; blank part dimensions:
; XA = 50 bar diameter
; ZA = 0.3 machining allowance on front face
; ZI = -100 length of finished part
; ZB = -70 extension from jaws
N10 WORKPIECE(,,,"CYLINDER",0,0.3,-100,-70,50)

N20 G18 G54 G90
N30 G0 X400 Z500
N40 M8
N50 SETMS(1)
```

```
N60 LIMS=2200
N70 T1 D1
N80 G96 S100 M4
→ N90 G0 X52 Z............ ; POSITIONING FOR FACING
N100 G1 X-1.6 F0.18
→ N110 G0 X............ Z0.5 ; STARTING DIAMETER OF CHAMFER AT 30°
→ N120 G1 X30 ANG=............ ; ANGLE FOR THE EXECUTION OF THE
CHAMFER
N130 G1 Z-12
→ N140 G1 X36 CHR=............ ; DIMENSION OF CHAMFER AT 45°
N150 G1 Z-18
→ N160 G1 Z-28 ANG=............ RND=............ ; INCLINATION ANGLE OF
FIRST LINE AND VALUE OF THE ROUND BETWEEN THE TWO SEGMENTS
→ N170 G1 X50 ANG=............ ; INCLINATION ANGLE OF SECOND LINE
N180 G1 Z-43
→ N190 G1 Z-59 ANG=............ ; INCLINATION ANGLE OF THE LINE
→ N200 G1 X50 RND=............ ; ROUND RADIUS WITH DIAM. 50
N210 G1 Z-61
N220 G1 X51

N230 G0 X200 ; DISENGAGEMENT
N240 G0 Z200
N250 M30
```

Compare your program to the one in the folder FINISHED_EXERCISES named EX_12_01.

13. Circular Interpolation (1h)
(Theory: 0.5h, Practice: 0.5h)

13.1 G2: circular interpolation in clockwise direction
The function G2 allows for the programming of circle arcs traveled by the tool in clockwise direction. The clockwise direction is defined according to the programming scheme shown in paragraph 4.9.
The circle arc is programmed by entering the function G2, followed by the coordinates of the arrival point and the radius dimensions (CR=).

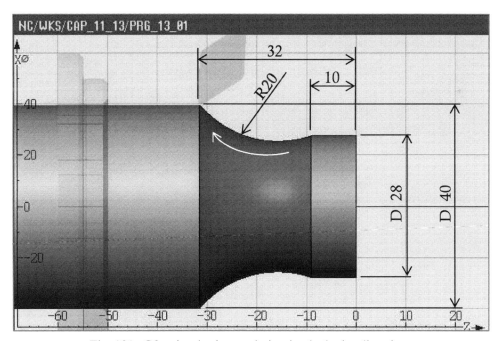

Fig. 101. G2 : circular interpolation in clockwise direction

Below you will find a programming example for the creation of the profile shown in figure 101:

```
N10 WORKPIECE(,,,"CYLINDER",0,0,-80,-50,40)
N20 G18 G54 G90
N30 G0 X400 Z500
N40 M8
N50 SETMS(1)
N60 T1 D1 ; TURNING TOOL
N70 G95 S1400 M4
N80 G0 X28 Z2
N90 G1 Z-10 F0.18
N100 G2 X40 Z-32 CR=20
N110 G1 X41
N120 G0 X200
N130 G0 Z200
N140 M30
```

13.2 G3: circular interpolation in counterclockwise direction

The function G3 allows for the programming of circle arcs traveled by the tool in counterclockwise direction.

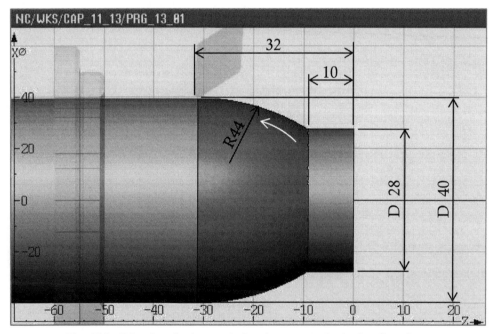

Fig. 102. G3 : circular interpolation in counterclockwise direction

Below you will find a programming example for the creation of the profile shown in figure 102:

```
N10  WORKPIECE(,,,"CYLINDER",0,0,-80,-50,40)
N20  G18 G54 G90
N30  G0 X400 Z500
N40  M8
N50  SETMS(1)
N60  T1 D1 ; TURNING TOOL
N70  G95 S1400 M4
N80  G0 X28 Z2
N90  G1 Z-10 F0.18
N100 G3 X40 Z-32 CR=44
N110 G1 X41
N120 G0 X200
N130 G0 Z200
N140 M30
```

13.3 I, K, I=AC(...), K=AC(...): progr. of the radius center

In the previous paragraphs, the circle arc has been defined by programming its arrival point and the radius value.

Another option is to program the coordinates of the radius center on X and Z (or Y when available) instead of the radius.

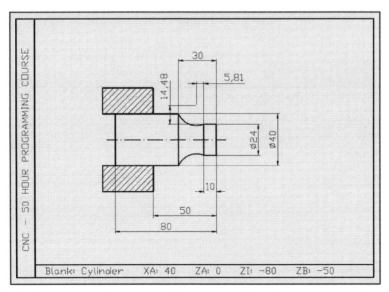

Fig. 103. Programming of an arc by means of the radius center coordinates

These coordinates may be expressed as incremental values referring to the starting point of the arc using the addresses I and K.

I: expresses the coordinate of the radius center with respect to the starting point of the arc on the X-axis (with radial value).
K: expresses the coordinate of the radius center with respect to the starting point of the arc on the Z-axis.

The following program executes the profile shown in figure 103:

```
N10  WORKPIECE(,,,"CYLINDER",0,0,-80,-50,40)
N20  G18 G54 G90
N30  G0 X400 Z500
N40  M8
N50  SETMS(1)
N60  T1 D1 ; TURNING TOOL
N70  G95 S1400 M4
N80  G0 X24 Z2
N90  G1 Z-10 F0.18
N100 G2 X40 Z-30 I14.48 K-5.81
N110 G1 X41
N120 G0 X200
N130 G0 Z200
N140 M30
```

It is also possible to program the absolute coordinates of the radius center referring to the part zero point using the following addresses:

I=AC(...), absolute coordinate in X (with diametral value) of the radius center on the X-axis.
K=AC(...): absolute coordinate in Z of the radius center on the Z-axis.

The same radius as in figure 103 can be programmed as follows:

```
N80  G0 X24 Z2
N90  G1 Z-10 F0.18
N100 G2 X40 Z-30 I=AC(52.96) K=AC(-15.81)
N110 G1 X41
```

J and J=AC(...) express the coordinate of the radius center with respect to the starting point of the arc on the Y-axis (plane G19).

13.4 Definition of the working plane

During turning operations, the tool always moves on the X-Z plane. When there is a Y-axis in the machine (see par. 4.5), there are two more working planes (X-Y and Z-Y), which are only used for milling operations.

The functions for circular interpolation require the programming of the working plane of the tool before their execution.

G18   defines the working plane X-Z
G19   defines the working plane Y-Z
G17   defines the working plane X-Y

In a lathe, the X-Z plane is normally already enabled at machine start. As long as turning operations are carried out, it is therefore unnecessary to reprogram it; amongst the first blocks the function G18 was entered again, only as a reminder of the presence of this instruction. The functions G17 and G19 are programmed before the milling operations performed on these planes.

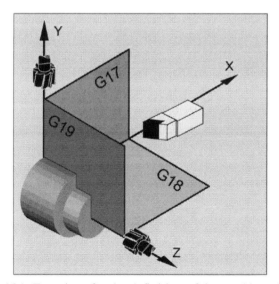

Fig. 104. Functions for the definition of the working plane

The definition of the working plane is essential, not only for the circular interpolations, but also for the tool radius compensation functions and for the direct programming of the angles.

13.5 Practical exercise

13.5.1 Programming of different radii
Open the program PRG_13_01 in the folder CHAP_11_13.
Copy it into the folder 01_EXERCISES and rename it as EX_13_01.
This program contains the execution of an arc in block N100.
Replace block N100 with the ones specified subsequently, start the graphic simulation, enable the single block mode and observe the programmed tool path.

In order to break the sharp edge at the end of the arc by means of a round, program RND= in the block of G2; remember to check and if necessary modify (as in this case) the direction of the next block to which it is joined.

```
N90 G1 Z-10 F0.18
N100 G2 X40 Z-32 CR=20 RND=4
G1 Z-36
N110 G1 X41
```

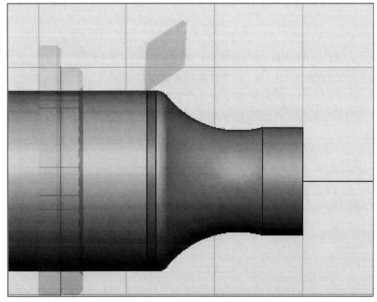

Fig. 105. Round between a radius and the following line by means of G2 and RND

In order to break the sharp edge at the end of the arc by means of a chamfer, program CHR= or CHF= in the block of G2 and, as before, remember to check the direction of the next block with which the arc is intersected.

```
N90  G1 Z-10 F0.18
N100 G2 X40 Z-32 CR=20 CHR=5
G1 Z-34
N110 G1 X41
```

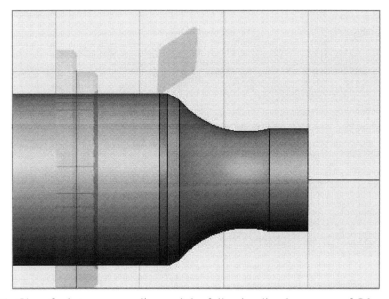

Fig. 106. Chamfer between a radius and the following line by means of G2 and CHR

Now use the function G3 and note how the shape of the programmed radius changes.

```
N90  G1 Z-10 F0.18
N100 G3 X40 Z-32 CR=40
N110 G1 X41
```

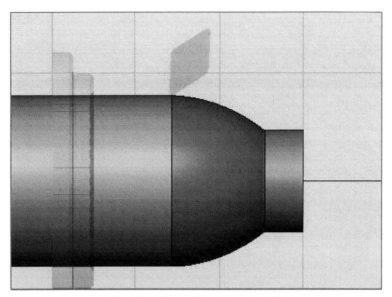

Fig. 107. Use of the function G3

Now replace block N100 with the following blocks and analyze the profile described by them.

```
N90  G1 Z-10 F0.18
N100 G2 X40 Z-32 I18.35 K-6.81
N110 G1 X41
```

```
N90  G1 Z-10 F0.18
N100 G2 X40 Z-32 I=AC(64.7) K=AC(-16.81)
N110 G1 X41
```

Try programming different radii by changing the coordinates of the arrival point and the radius dimensions. If there are radii and points which are not correctly programmed, alarms will be displayed.

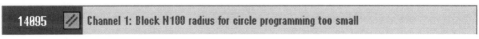

Fig. 108. Type of alarm displayed in the event of a programming error in a radius

14. First Test (2h)
(Practice: 2h)

14.1 Introduction to the test

The test consists in the execution of the program creating the part shown in figure 111. Take the following steps:
- Load the tool files contained in folder 01_EXERCISES named EMPTY_TOOL_LIST. This file deletes all the existing tools by overwriting them only with the roughing tool defined therein. In order to load this file, follow the procedure laid down in paragraph 3.3.
- Now create the necessary tools for the execution of this program following the procedure specified in paragraph 7.5.1. Below is a list of the necessary tools, their position in the turret, the offset data in X and Z and the data for the definition of their graphic aspect.

Loc.	Type	Tool name	ST	D	Length X	Length Z	Ø	Tip angle		
1		ROUGHING TOOL	1	1	88.000	40.000	0.800 ←	93.0	55	11.0
2		FINISHING TOOL	1	1	94.000	40.000	0.200 ←	93.0	55	11.0
3		OD GROOVING W.3MM	1	1	98.000	40.000	0.100	3.000		10.0
4		OD THREADING	1	1	88.000	46.000	0.200			
5		CENTER DRILL D.6	1	1	100.000	24.000	6.000	118.0		
6		AX. DRILL D.8.5	1	1	100.000	56.000	8.500	118.0		

Fig. 109. List of tools to be created and used in the test program

- Create an empty main program in the folder 01_EXERCISES and name it TEST_14_01.
- Structure the program as the ones we've seen so far:

- Enter the comments with the dimensions of the blank part at the beginning of the program. If you want to copy blocks from already existing programs see paragraph 14.4.
- Define the dimensions of the blank part according to the procedure laid down in paragraph 12.6.2.
- Enter the blocks that activate the initial settings and the safety position:
  ```
  G18 G54 G90
  G0 X400 Z500
  M8
  SETMS(1)
  ```
- Proceed to the programming of the tooling operations following the logical sequence described in paragraph 5.2.

14.2 Tooling operations and cutting parameters

Tooling sequence	Tool	Operation	Cutting speed (m/min)	Feedrate (mm/rev)
1 st	T1 D1	Roughing	100	0.18
2 nd	T2 D1	Finishing	120	0.12
3 rd	T3 D1	Groove	78	0.1
4 th	T4 D1	Threading	60	-
5 th	T5 D1	Center drilling	80	0.08
6 th	T6 D1	Hole D8.5	80	0.1

Fig. 110. Sequence of tooling operations and cutting parameters to use in the test

14.3 Drawing of the part to create

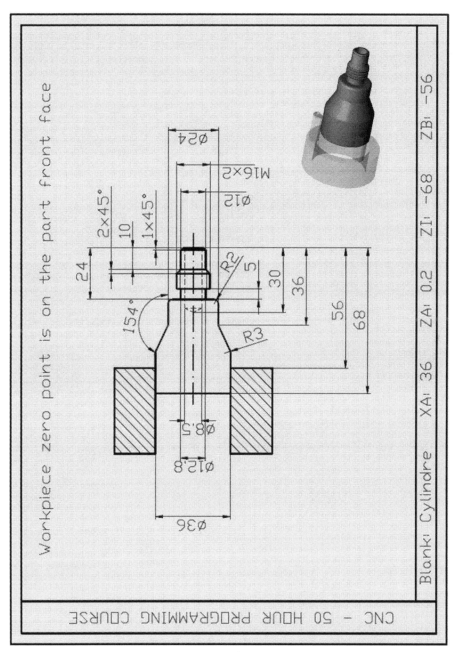

Fig. 111. Drawing of the part to create

14.4 Copying & pasting of program parts

In order to speed up programming times it is possible to copy or cut part of a program and paste it into a new position.

Place the cursor on the initial block of the program part you want to copy or cut,

then press MARK,

go down with the cursor and select the blocks.

Then press COPY or CUT,

place the cursor where you want to copy or move the selected text. The new position may be in the same program or in another program.

Then press PASTE.

14.5 Program correction

Compare your program to the one in the folder FINISHED_EXERCISES named TEST_14_01.

15. Tool Radius Compensation (1h)
(Theory: 0.5h, Practice: 0.5h)

15.1 Introduction

When offsetting a tool, the distance between the tip of the tool and the characteristic point of the slide on all the axes on which the slide moves (in this case X and Z), is entered into the specific geometry page (paragraph 6.2.2).

Whether these values are obtained by touching the workpiece or by measuring them outside of the machine, the point defined on the X-axis does not correspond to the point defined on the Z-axis.

This is due to the presence of the insert radius.

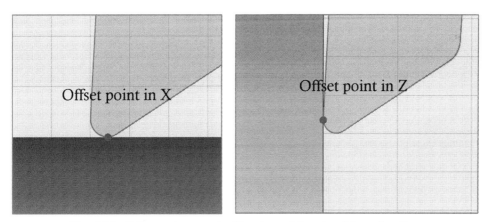

Fig. 112. Offset points on the X-axis and Z-axis with tool radius

The distance between the two points increases when the dimension of the insert radius increases.

The offset values in X and Z define the coordinates of the point used by the NC in order to execute the programmed path. This is on the cutting edge without considering the presence of the insert radius, as if the tool had a sharp edge (see fig. 113).

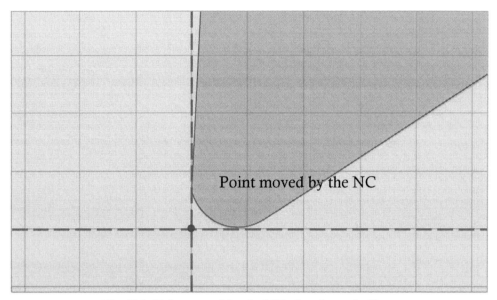

Fig. 113. Point moved by the NC after the tool offset

When executing cylindrical turnings and vertical shoulderings, the presence of the insert radius does not lead to changes in the execution of the profile, as the point moved by the NC is located exactly on the cutting edge which determines the shape and the dimension of the workpiece.

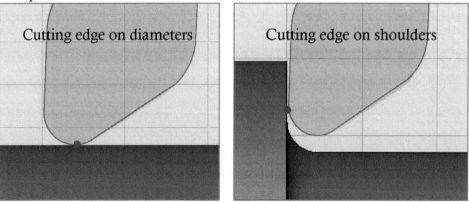

Fig. 114. Absence of changes due to the tool radius on diameters and shoulders

The presence of the tool radius means errors in the description of the profile during the turning of conical workpieces and in the execution of circular interpolations.

When the tool travels along conical profiles, the point moved by the NC does not correspond to the edge of the tool cutting the material. This leads to an error in the dimensions of the described profile.
The inclination angle does not change and keeps the created conical shape geometrically correct.

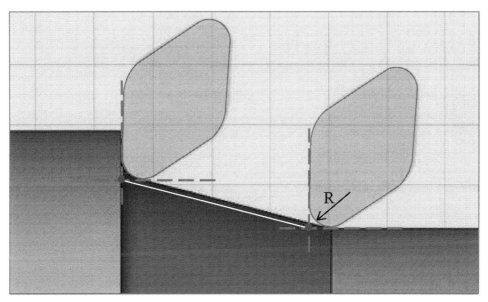

Fig. 115. Dimensional error caused by the insert radius during the execution of conical turnings

As you can see in the figure, the programmed profile, represented by the white line, does not correspond the profile created by the tool.

Also in the execution of circular interpolations, the described profile does not correspond to the programmed profile.
In the following figure you can see how different the programmed profile is from the one actually created by the tool.

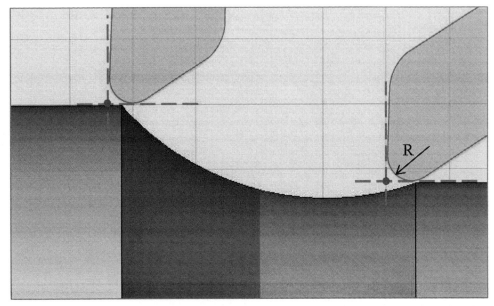

Fig. 116. Error caused by the insert radius during the execution of a circular interpolation

The automatic correction of the tool path is performed by enabling the modal functions G42 and G41; these functions are disabled by the function G40.

The information necessary for the NC for the automatic correction of the tool path are:
- dimension of the tool radius
- position of the radius with respect to the zero point

This information is entered in the geometry page during the graphic description of the tool (s. fig. 67 and 68 in chapter 7).

In all the NCs that do not have a graphic description of the tool, the orientation of the zero point compared to the radius (also called the quadrant of the tool) is defined according to an ISO Standard code shown in the following figure.

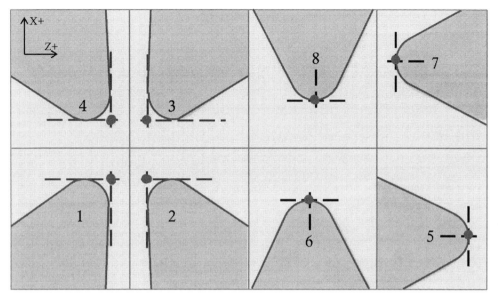

Fig. 117. Quadrant code defining the radius position with respect to the zero point

When the zero point is at the center of the radius which needs to be compensated (as with mills), the quadrant code defining the radius must be 0 or 9.

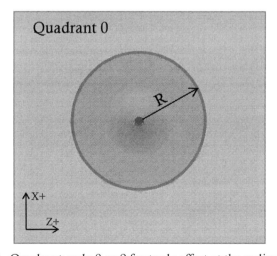

Fig. 118. Quadrant code 0 or 9 for tools offset at the radius center

15.2 G42: Enabling with tool on right side of profile

After the definition of the radius value and its position with respect to the zero point in the geometry table, it is necessary to enable the appropriate function for the tool radius compensation in the program.

When the tool is located on the right side of the profile, the function G42 is used. The right and the left side are defined by the cutting direction of the tool.

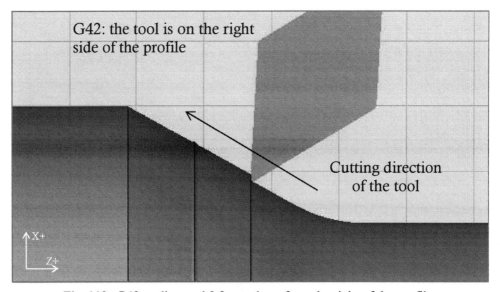

Fig. 119. G42: radius tool 0.2, quadrant 3, to the right of the profile

Attention! The right and the left side are determined as if you were walking on the profile in cutting direction.

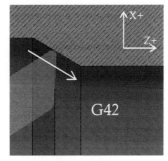

Fig. 120. G42: radius tool 0.8, quadrant 1, to the right of the profile

15.3 G41: Enabling with tool on left side of profile

With the tool on the left side of the program, the function G41 is programmed.

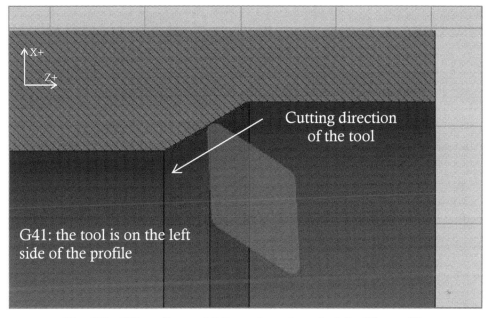

Fig. 121. G41: radius tool 0.8, quadrant 2, to the left of the profile

Make sure not to associate the functions G42 or G41 with external or internal profiles, as **the only correct assessment is to recognize the position of the tool to the right or to the left of the profile with respect to the cutting direction.**

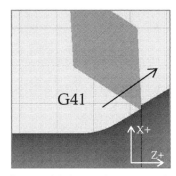

Fig. 122. G41: radius tool 0.2, quadrant 4, to the left of the profile

15.4 Enabling and disabling with G40

In the blocks with the enabling and disabling functions for the tool radius compensation, the NC corrects the programmed path in order to be prepared to execute the profile in a correct manner.

It is therefore necessary to always enter these functions in blocks outside of the profile, as for example the approaching block for enabling and the disengaging block for disabling. Make sure that the programmed offset is higher than the dimension of the radius to compensate.

G40 is the function that disables G41 and G42.

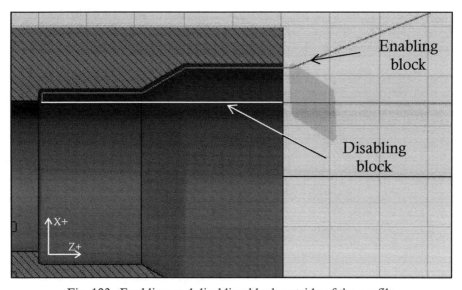

Fig. 123. Enabling and disabling block outside of the profile

Below you will find a programming example for the definition of the profile shown in the figure above.

```
T12 D1
G95 S1800 M4
G0 X45 Z2 G41 ; ENABLING BLOCK OUTSIDE OF PROFILE
G1 Z-20 F0.1
G1 X35 ANG=210
G1 Z-50
G1 X28
G0 Z2 G40 ; DISABLING BLOCK OUTSIDE OF PROFILE
G0 Z200 X200
```

15.5 Practical exercise

15.5.1 Program analysis
Enter the missing data into the program below.
The arrow (→) before the block number tells you where to enter the missing value.

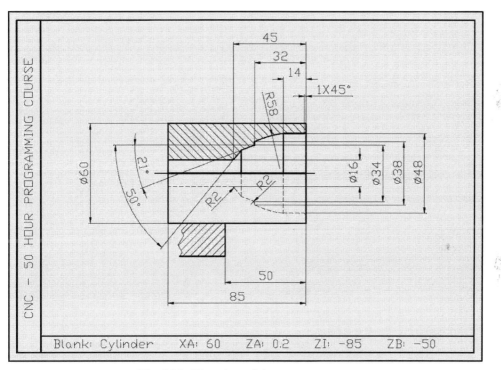

Fig. 124. Drawing of the part to create

```
; blank part dimensions:
; XA = 60 bar diameter
; ZA = 0.2 machining allowance on front face
; ZI = -85 length of finished part
; ZB = -50 extension from jaws
N10 WORKPIECE(,,,"CYLINDER",0,0.2,-85,-50,60)

N20 G18 G54 G90
N30 G0 X400 Z500
N40 M8
N50 SETMS(1)

N60 LIMS=3000 ; LIMITATION TO 3000 REV/MIN
```

```
N70 T1 D1 ; TURNING TOOL FOR EXTERNAL PARTS
N80 G96 S100 M4 ; ENABLING OF CONSTANT CUTTING SPEED

N90 G0 X62 Z0 ; APPROACH
→ N100 G1 X............... F0.18 ; FACING
N110 G0 X200 Z200 ; DISENGAGEMENT

N120 T11 D1 ; RIGHT AXIAL DRILL DIAMETER 16 MM
N130 G95 S1100 M3 ; ENABLING OF FIXED NUMBER OF REVOLUTIONS
N140 G0 X0 Z2 ; APPROACH
N150 G1 Z-30 F0.12 ; FIRST DRILLING PASS
N160 G4 S2 ; DWELL TIME OF 2 SPINDLE REVOLUTIONS
N170 G0 Z5 ; RAPID EXIT FOR CHIP REMOVAL
N180 G1 Z-29 F2 ; ENTERING WITH HIGH FEEDRATE
N190 G1 Z-60 F0.12 ; SECOND PASS UP TO Z-60
N200 G4 S2
N210 G0 Z5
→ N220 G1 Z............... F2
N230 G1 Z-90 F0.12
N240 G0 Z200 ; DISENGAGEMENT IN Z

N250 T12 D1 ; BORING BAR FOR INTERNAL TURNING
N260 G96 S120 M4 ; ENABLING OF CONSTANT CUTTING SPEED

N270 G0 X22 Z2
N280 G1 Z-41 F0.14 ; FIRST ROUGHING PASS
N290 G0 X20 Z2
N300 G0 X28
N310 G1 Z-34 ; SECOND ROUGHING PASS
N320 G0 X26 Z2
N330 G0 X34
N340 G1 Z-31.8 ; THIRD ROUGHING PASS
N350 G0 X32 Z2
N360 G0 X40
N370 G1 Z-28 ; FOURTH ROUGHING PASS
N380 G0 X38 Z2
N390 G0 X46
N400 G1 Z-16 ; FIFTH ROUGHING PASS
N410 G0 X44 Z5

;BEGINNING OF FINISHING WITH SAME TOOL
N420 G96 S150 M4 ; ENABLING OF CONSTANT CUTTING SPEED

N430 G0 X50 Z5
N440 G0 Z2 G41
N450 G1 Z0 F0.1
N460 G1 Z-1 ANG=225
```

→ N470 G1 Z............
→ N480 G3 X............ Z-32 CR=58
→ N490 G1 X34 RND=............
N500 G1 ANG=201
→ N510 G1 X............ Z-45 ANG=230 RND=2
N520 G1 Z-48
N530 G1 X15
N540 G0 Z5 G40
N550 G0 X200 Z200
N560 M30

Now open the program EX_15_01 in the folder 01_EXERCISES and enter the values.

Before starting the graphic simulation create both of the tools used in this cycle (paragraph 7.5):
- in position 11, an axial drill diameter 16 mm
- in position 12, a boring bar.

Use the names and data shown in the figure below.

11	AX. DRILL D.16	1	1	100.000	120.000	16.000		118.0		
12	ROUGH. BORING-BAR	1	1	86.000	92.000	0.400 ←		93.0	55	8.0

Fig. 125. Data of the new tools to create for the execution of the cycle

If there are problems with the creation of the new tools, you can review the practical exercise in paragraph 7.5, or proceed by loading the tool file named TOOL_LIST from the folder 01_EXERCISES according to the procedures laid down in paragraph 3.3.

Start the graphic simulation in single block mode and display the HALF CUT VIEW of the workpiece, analyze the program and change it where necessary.

Compare your program to the one in the folder FINISHED_EXERCISES named EX_15_01.

15.5.2 Test of concept comprehension

The following exercise proposes different combinations of tool radius compensation enabling function and radius quadrant code.
Put a cross by the only correct combination among the given answers.

1)

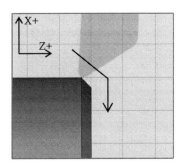

a) G42, quadrant 3

b) G41, quadrant 3

c) G41, quadrant 4

2)

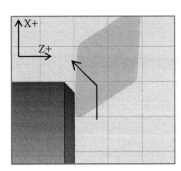

a) G41, quadrant 1

b) G42, quadrant 3

c) G42, quadrant 2

3)

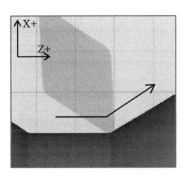

a) G42, quadrant 4

b) G42, quadrant 2

c) G41, quadrant 4

161

4)

a) G41, quadrant 2

b) G42, quadrant 2

c) G41, quadrant 3

5)

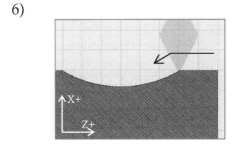

a) G42, quadrant 2

b) G41, quadrant 2

c) G42, quadrant 1

6)

a) G41, quadrant 6

b) G42, quadrant 3

c) G42, quadrant 8

The correct solutions can be found in the program ANS_15_01 in the folder CHAP_15_20.

15.6 Reloading of complete tool list

Before reading the next chapter, reload the complete tool list contained in the folder 01_EXERCISES.

16. Advanced Programming Functions (2h)
(Theory: 1h, Practice: 1h)

16.1 Introduction
The functions seen so far allow for the writing of the complete programs of all the parts which can be created in a lathe with 2 axes (X and Z).
From this chapter onwards useful methods for the optimization of programming times will be specified, sometimes involving a shortening of the program via the call of subprograms and by skipping or repeating program parts, at others by letting the NC calculate the movements to be executed by the tool, becoming familiar with how to use fixed working cycles.

16.2 Call of a subprogram
The creation of a subprogram has been explained in paragraph 8.10.3.
The call of a subprogram is recommended when the same profile is to be repeated frequently or there is to be repetition of an identical sequence of movements, such as with a series of grooves or the return to the home position. This helps to shorten the main program and to simplify a possible correction of the profile, which will be edited only in the subprogram and not in all the blocks of the main program, where the profile would otherwise be repeated.
The structure of a subprogram is identical to the structure of a main program and its ideal placement is in the same workpiece folder where the main program is located, or in the specific subprogram folder (s. paragraph 3.2).
In order to call a subprogram it is sufficient to write its name in a block without any other function.
The function M17 (or RET) closes the subprogram and causes it to return to the main program after the calling block.
The following main program calls the subprogram named as 'GROOVE' in order to create a part characterized by the presence of multiple grooves with an identical profile.

The main program chooses the tool, sets the rotation speed of the spindles and the feedrate type used, positions the tool in absolute coordinates on the starting point of the first groove and calls the appropriate subprogram to describe its profile.

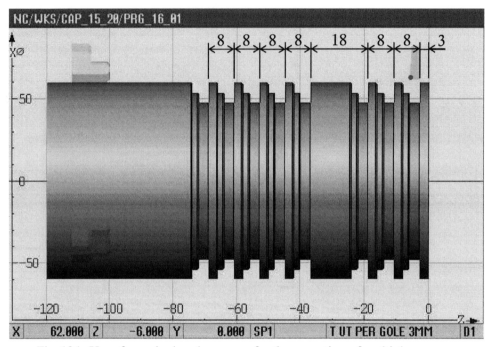

Fig. 126. Use of one single subprogram for the execution of multiple grooves

```
; blank part dimensions:
; XA = 60 bar diameter
; ZA = 0 machining allowance on front face
; ZI = -120 length of finished part
; ZB = -100 extension from jaws
WORKPIECE(,,,"CYLINDER",0,0,-120,-100,60)

G18 G54 G90
G0 X400 Z500
M8
SETMS(1)
```

T3 D1 ; 3 MM GROOVING TOOL WITH CUTTING EDGE OFFSET ON THE LEFT
G95 S1150 M4 ; SETTING OF NUMBER OF REVOLUTIONS AND FEEDRATE IN MM/REV

```
G0 X62 Z-6 ; POS. IN ABSOLUTE COORDINATES OF THE STARTING
POINT OF THE FIRST GROOVE
GROOVE ; CALL OF SUBPROGRAM

G0 Z-14 ; STARTING POINT OF SECOND GROOVE
GROOVE

G0 Z-22 ; STARTING POINT OF THIRD GROOVE
GROOVE

G0 Z-40 ; STARTING POINT OF FOURTH GROOVE
GROOVE

G0 Z-48 ; STARTING POINT OF FIFTH GROOVE
GROOVE

G0 Z-56 ; STARTING POINT OF SIXTH GROOVE
GROOVE

G0 Z-64 ; STARTING POINT OF SEVENTH GROOVE
GROOVE

G0 Z-72 ; STARTING POINT OF EIGTH GROOVE
GROOVE

G0 X200
G0 Z200

M30
```

The following subprogram named GROOVE.SPF contains the profile of the groove. Note how in this case the mixed programming, which allows for the expression of some coordinates in the absolute reference system and others in the incremental reference system (see chapter 10), is essential at the moment when the profile needs to be repeated in different points of the workpiece.

It is obvious that thanks to the use of a subprogram, the main program becomes shorter and easier to read. Furthermore, if the profile needs to be changed, the correction of one single subprogram guarantees the perfect replication of the tooling operation to be carried out.

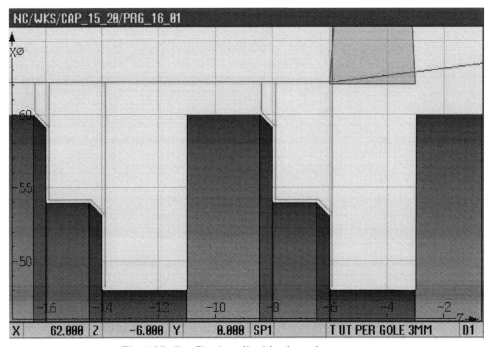

Fig. 127. Profile described in the subprogram

```
G1 X48.1 F0.12 ; ROUGHING OF THE GROOVE AT DIAMETER 48.1
G4 S2 ; DWELL TIME OF TWO SPINDLE REVOLUTIONS
G1 X62 F1 ; RETURN TO DIAMETER 62 WITH HIGH FEEDRATE

G1 Z=IC(-2) ; REPOSITIONING IN INCREMENTAL SYSTEM
G1 X54.1 F0.12 ; ROUGHING OF THE GROOVE AT DIAMETER 54.1
G4 S2
G1 X62 F1

;EXECUTION OF THE CHAMFERS AND FINISHING OF THE PROFILE
G1 Z=IC(-0.5)
G1 X60 F0.12
G1 Z=IC(0.5) ANG=-45
G1 X54
G1 Z=IC(1.5)
G1 Z=IC(0.5) ANG=-45
G1 X48
G1 X62 F1
```

M17 ; CLOSING FUNCTION OF SUBPROGRAM

16.3 Repetition of a subprogram

When the elements to be created are characterized by a constant shift (e.g. grooves formed at constant lead or equidistant holes), it is possible to repeat the execution of a subprogram which, at its last block, will contain the repositioning in incremental coordinates at the starting point of the next tooling operation.

Below you will find a programming alternative for the creation of the part shown in figure 126.

The first three grooves are at a fixed distance of 8 mm, then there is a repositioning in absolute coordinates for the execution of the second groove group, which are also equidistant by 8 mm.

```
; blank part dimensions:
; XA = 60 bar diameter
; ZA = 0 machining allowance on front face
; ZI = -120 length of finished part
; ZB = -100 extension from jaws
WORKPIECE(,,,"CYLINDER",0,0,-120,-100,60)

G18 G54 G90
G0 X400 Z500
M8
SETMS(1)

T3 D1 ; 3MM GROOVING TOOL WITH CUTTING EDGE OFFSET ON THE LEFT
G95 S1150 M4

G0 X62 Z-6 ; POS. IN ABSOLUTE COORDINATES OF THE STARTING POINT OF THE FIRST GROOVE
GROOVE_2E P=3 ; CALL OF THE SUBPROGRAM AND REPETITION FOR 3 TIMES

G0 Z-40 ; STARTING POINT OF FOURTH GROOVE
GROOVE_2E P=5 ; CALL OF THE SUBPROGRAM AND REPETITION FOR 5 TIMES

G0 X200
G0 Z200

M30
```

The new subprogram contains as last block the incremental shift to the next groove.

```
G1 X48.1 F0.12 ; ROUGHING OF THE GROOVE AT DIAMETER 48.1
G4 S2 ; DWELL TIME OF TWO SPINDLE REVOLUTIONS

G1 X62 F1 ; RETURN TO DIAMETER 62 WITH HIGH FEEDRATE

G1 Z=IC(-2) ; REPOSITIONING IN INCREMENTAL SYSTEM
G1 X54.1 F0.12 ; ROUGHING OF THE GROOVE AT DIAMETER 54.1
G4 S2

G1 X62 F1

;EXECUTION OF THE CHAMFERS AND FINISHING OF THE PROFILE
G1 Z=IC(-0.5)
G1 X60 F0.12
G1 Z=IC(0.5) ANG=-45
G1 X54
G1 Z=IC(1.5)
G1 Z=IC(0.5) ANG=-45
G1 X48
G1 X62 F1
G1 Z=IC(-8)

M17 ; CLOSING FUNCTION OF SUBPROGRAM
```

16.4 Labels: references for the setting of skips

Labels are single words written in a block which identify a specific position in the program.

They are used by the programmer to skip or repeat parts in the program.

The name of the label can have from a minimum of 2 to a maximum of 32 alphanumeric characters (the first two characters need to be letters or underscores).

Their names must not be the same as the name of a function (e.g. GOTO, LOOP) and they always need to be programmed followed by a colon (:).

The colon at the end of a label is essential to distinguish it from the call of a subprogram.

In the following paragraphs we will see how to use them.

16.5 REPEAT: repetition of blocks

If you want to repeat the same sequence of blocks you can use the REPEAT function.

A label is entered in order to mark the beginning of the block to be repeated. Then the block sequence is entered and then the REPEAT function is used, followed by the name of the label and by the number of repetitions shown by the address 'P='.

```
...
GROOVE_START:
G1 X48.1 F0.12 ; ROUGHING OF THE GROOVE AT DIAMETER 48.1
G4 S2 ; DWELL TIME OF TWO SPINDLE REVOLUTIONS

G1 X62 F1 ; RETURN TO DIAMETER 62 WITH HIGH FEEDRATE

G1 Z=IC(-2) ; REPOSITIONING IN INCREMENTAL SYSTEM
G1 X54.1 F0.12 ; ROUGHING OF THE GROOVE AT DIAMETER 54.1
G4 S2

G1 X62 F1

;EXECUTION OF THE CHAMFERS AND FINISHING OF THE PROFILE
G1 Z=IC(-0.5)
G1 X60 F0.12
G1 Z=IC(0.5) ANG=-45
G1 X54
G1 Z=IC(1.5)
G1 Z=IC(0.5) ANG=-45
G1 X48
G1 X62 F1
G1 Z=IC(-8)
REPEAT GROOVE_START P=3
...
```

Another option is to define the profile within two labels specifying the beginning and the end; this allows for the repetition of the sequence of blocks in any part of the program.

Below you will find a programming alternative for the creation of the same part shown in figure 126 using the REPEAT function.

```
; blank part dimensions:
; XA = 60 bar diameter
; ZA = 0 machining allowance on front face
; ZI = -120 length of finished part
; ZB = -100 extension from jaws
WORKPIECE(,,,"CYLINDER",0,0,-120,-100,60)

G18 G54 G90
G0 X400 Z500
M8
SETMS(1)

T3 D1 ; 3 MM GROOVING TOOL WITH CUTTING EDGE OFFSET ON THE LEFT
G95 S1150 M4

G0 X62 Z-6 ; POS. IN ABSOLUTE COORDINATES OF THE STARTING POINT OF THE FIRST GROOVE

GROOVE_START:
G1 X48.1 F0.12 ; ROUGHING OF THE GROOVE AT DIAMETER 48.1
G4 S2 ; DWELL TIME OF TWO SPINDLE REVOLUTIONS

G1 X62 F1 ; RETURN TO DIAMETER 62 WITH HIGH FEEDRATE

G1 Z=IC(-2) ; REPOSITIONING IN INCREMENTAL SYSTEM
G1 X54.1 F0.12 ; ROUGHING OF THE GROOVE AT DIAMETER 54.1
G4 S2

G1 X62 F1

;EXECUTION OF THE CHAMFERS AND FINISHING OF THE PROFILE
G1 Z=IC(-0.5)
G1 X60 F0.12
G1 Z=IC(0.5) ANG=-45
G1 X54
G1 Z=IC(1.5)
G1 Z=IC(0.5) ANG=-45
G1 X48
G1 X62 F1
G1 Z=IC(-8)
GROOVE_END:

REPEAT GROOVE_START GROOVE_END P=2
```

```
G0 Z-40 ; STARTING POINT OF FOURTH GROOVE
```

REPEAT GROOVE_START GROOVE_END P=5

```
G0 X200
G0 Z200

M30
```

The function REPEAT offers a useful alternative to the use of the subprogram, allowing for the programming of the sequence of blocks to be repeated in the main program, thereby simplifying the procedures for the sequence modification.

16.6 GOTO: skipping parts of a program

The GOTO function allows one to skip directly to a selected label.
It may be used to skip program parts which are temporarily not being used in the machine tooling phase, or to skip profiles defined at the beginning of the program but used only later; often programmers use this programming style to make the profiles of the workpiece easily recognizable when modifying the program.

```
; blank part dimensions:
; XA = 60 bar diameter
; ZA = 0 machining allowance on front face
; ZI = -120 length of finished part
; ZB = -100 extension from jaws
WORKPIECE(,,,"CYLINDER",0,0,-120,-100,60)
```

GOTO IN_PROGRAM

GROOVE_START:
```
G1 X48.1 F0.12 ; ROUGHING OF THE GROOVE AT DIAMETER 48.1
G4 S2 ; DWELL TIME OF TWO SPINDLE REVOLUTIONS

G1 X62 F1 ; RETURN TO DIAMETER 62 WITH HIGH FEEDRATE

G1 Z=IC(-2) ; REPOSITIONING IN INCREMENTAL SYSTEM
G1 X54.1 F0.12 ; ROUGHING OF THE GROOVE AT DIAMETER 54.1
G4 S2

G1 X62 F1

;EXECUTION OF THE CHAMFERS AND FINISHING OF THE PROFILE
```

```
G1 Z=IC(-0.5)
G1 X60 F0.12
G1 Z=IC(0.5) ANG=-45
G1 X54
G1 Z=IC(1.5)
G1 Z=IC(0.5) ANG=-45
G1 X48
G1 X62 F1
G1 Z=IC(-8)
```
GROOVE_END:

IN_PROGRAM:

```
G18 G54 G90
G0 X400 Z500
M8
SETMS(1)
```

T3 D1 ; 3 MM GROOVING TOOL WITH CUTTING EDGE OFFSET ON THE LEFT
G95 S1150 M4

G0 X62 Z-6 ; POS. IN ABSOLUTE COORDINATES OF THE STARTING POINT OF THE FIRST GROOVE

REPEAT GROOVE_START GROOVE_END P=3

G0 Z-40 ; STARTING POINT OF FOURTH GROOVE

REPEAT GROOVE_START GROOVE_END P=5

```
G0 X200
G0 Z200

M30
```

16.7 Conclusions

In the previous paragraphs we have seen four different programming methods which all allow for the creation of the same workpiece as shown in fig. 126.

Among methods, it is not possible to say which is best. Advanced programming offers broad alternatives for the writing of a program, whose features may be more or less appreciated subjectively by the programmer.

The only basic rule is that of creating a program in the manner that you consider most simple and understandable, always bearing in mind that the operator of the machine is not necessarily a programming expert.

16.8 Practical exercise

16.8.1 Call of a subprogram
Analyze the program described in paragraph 16.2.
Open the program 'PRG_16_01' in the folder 'CHAP_15_20'
The main program creates the part shown in figure 126 by calling the subprogram GROOVE.SPF multiple times.
Activate the graphic simulation in single block mode, then analyze the structure of the program and the values for the tool positioning before the call of the subprogram.
Furthermore, analyze the use of the mixed programming used in the subprogram for the definition of the profile of the groove.

16.8.2 Repetition of a subprogram
Now analyze the program described in paragraph 16.3.
Open the program 'PRG_16_02'.
This time the main program repeats the execution of the subprogram GROOVE_2E.SPF, where the repositioning in incremental coordinates for the execution of the following tooling operation was added as last block.

16.8.3 Use of the REPEAT function
Now analyze the second programming example described in paragraph 16.5. **Open the program 'PRG_16_03'.**
The main program does not call a subprogram, but repeats a certain block sequence defined by the start and end label.

16.8.4 Use of the GOTO function
Analyze the program described in paragraph 16.6.
Open the program 'PRG_16_04'.
Now ask yourselves why the program PRG_16_03, for the execution of the first three grooves, repeats the profile only twice while the current program repeats the profile three times.

```
REPEAT GROOVE_START GROOVE_END P=3
```

The answer can be found in the program ANS_16_04 in the folder CHAP_15_20.

17. Fixed Cycles (1h)
(Theory: 0.5h, Practice: 0.5h)

17.1 Introduction
A fixed cycle (also known as canned cycle) is a macroinstruction which, combined with a series of parameters, is able to execute a complete specific tooling operation autonomously.
As already seen in paragraph 11.5.1, the programming of the roughing of a piece, executed by means of ISO functions, is carried out by entering a long series of blocks. The greatest difficulty is the calculation of the points defining the tool path and their possible subsequent modification.
The fixed cycles serve to speed up and simplify the programming by controlling all the tool movements on the basis of the value of the associated parameters.
The cycles are developed by the various manufacturers of numeric controls; Siemens proposes a dialog box explaining the meaning of the parameters for every cycle. After its insertion in the program, the cycle has the shape of a block beginning with the name of the cycle and continuing with a sequence of numbers and letters.

```
N80 G0 X82 Z2¶
N90 G1 X6 F0.2¶
N100 G0 X80 Z2¶
¶
N110 CYCLE62(,2,"PROFILE1","END1")¶
¶
N120 CYCLE952("con_temp",,"",2301311,0.1,0,0,3,0.1,0.1,0.5,0.1,0.1,0
¶
N130 G0 X200 Z200¶
¶
```

Fig. 128. Insertion of fixed cycles in a program

In order to change a cycle after its insertion select the corresponding block with the cursor and **press the right arrow button** on the control panel or press directly the arrow shown at the end of the block with the pointer.

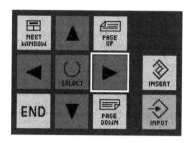

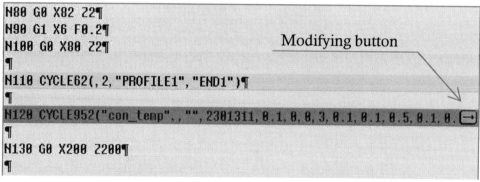

Fig. 129. Button for the modification of a cycle after its insertion

It must be noted that every tooling operation can always be programmed with ISO functions and that the cycles are only a useful alternative option.

17.2 Use of the HELP button

The HELP button is located on the control panel next to the arrow buttons. When pressing the HELP button, the NC recognizes the cursor position and searches, among its internal documents, for a description of what the operator has selected.

This button gives information about the commands present in the program, about the alarms and it's also particularly useful whenever short and immediate help in relation to the meaning of the parameters to be entered in the cycles is needed.

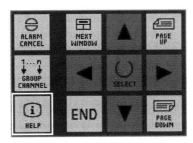

Fig. 130. HELP button on the control panel

In order to display the help window regarding the parameters contained in the cycles, open the cycle as previously shown and push the HELP button. A list of parameters with their descriptions appears. In order to close the help window press EXIT HELP.

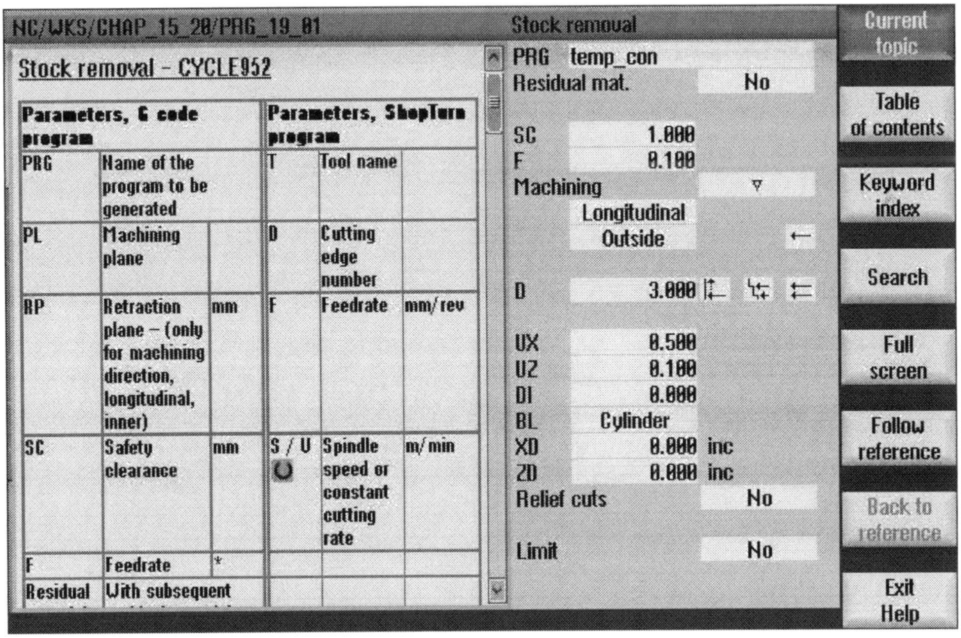

Fig. 131. Help window with description of cycle parameters

17.3 Insertion of a cycle in a program

In order to insert a cycle in a program, follow the steps below:
- open the program,
- place the cursor on the insertion block,
- starting from the main menu shown in the figure below, follow the procedure specified at the beginning of the following chapters.

Fig. 132. Start menu for the insertion of the fixed working cycles

17.4 Deletion of a cycle from the program

In order to delete a block containing a cycle, follow the steps below:
- press the horizontal softkey EDIT
- place the cursor on the block you want to delete
- press the vertical softkey CUT (if not displayed press the vertical softkey BACK)

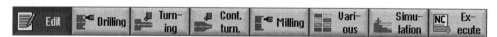

Fig. 133. Return to the EDIT menu to delete the cycles

17.5 Practical exercise

17.5.1 Buttons for cycle management

Open the program 'PRG_17_01' in the folder 'CHAP_15_20'. Try to use the button for the modification and reopening of the cycle, the HELP button in order to display the online help offered by the NC and delete the cycles inserted in the program by using the notions learned in this chapter.

Modify, delete and insert elements in this program until you're confident in the use of the buttons for cycle management.

18. CYCLE62: Profile Selection (1h)
(Theory: 0.5h, Practice: 0.5h)

18.1 Description
Before activating the roughing cycle it is necessary to identify the profile to be roughed using the cycle CYCLE62.

The cycle offers various options for the positioning of the profile:
- in the main program and defined between two labels,
- in a subprogram that contains only the profile to be roughed,
- in a subprogram where the profile is defined between two labels,
- in a main program, defining the tool path by means of the graphic profile generator contained in the numeric control.

18.2 Insertion procedure
Starting from the main menu shown in the paragraph 17.3, proceed by pressing the buttons shown in the following table.

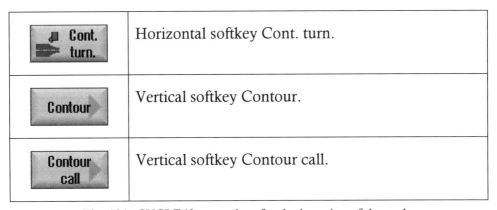

Cont. turn.	Horizontal softkey Cont. turn.
Contour	Vertical softkey Contour.
Contour call	Vertical softkey Contour call.

Fig. 134. CYCLE62: procedure for the insertion of the cycle

18.3 Parameters

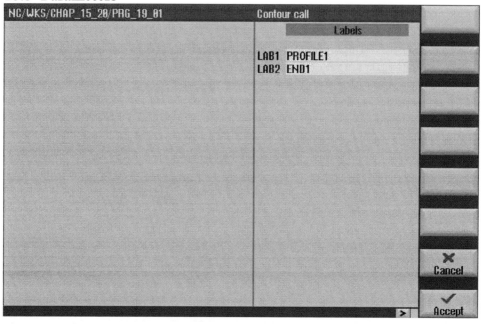

Fig. 135. CYCLE62: window for the insertion of the parameters

Option 1: labels

Parameter	Description
Labels	Select this option with the SELECT button; the profile is programmed in the main program; the beginning and the end of the block sequence is defined by the use of two labels.
LAB1	Name of the label for profile start (e.g. PROFILE1).
LAB2	Name of the label for profile end (e.g. END1).

Fig. 136. CYCLE62: label option

Option 2: subprogram

Parameter	Description
Subprogram	Select this option with the SELECT button; the profile is programmed in a subprogram located in the workpiece directory or in the SUBPROGRAMS folder.
PRG	Name of the subprogram containing the profile (example for the name of the subprogram PROFILE1.SPF).

Fig. 137. CYCLE62: subprogram option

Option 3: label in subprogram

Parameter	Description
Label in sub-program	Select this option with the SELECT button; the profile is programmed in a subprogram and limited by two labels for the beginning and the end.
PRG	Name of the subprogram containing the profile.
LAB1	Name of the label for profile start.
LAB2	Name of the label for profile end.

Fig. 138. CYCLE62: label in subprogram option

Option 4: profile name

Parameter	Description
Profile name	Select this option with the SELECT button; the profile is defined by means of the graphic profile generator.
CON	Name of the saved contour.

Fig. 139. CYCLE62: profile name option

The graphic profile generator offers the possibility to describe the contour of the workpiece with graphic buttons.
We will not go into detail for this option as it is not in line with the goal to learn how to use the ISO Standard functions.

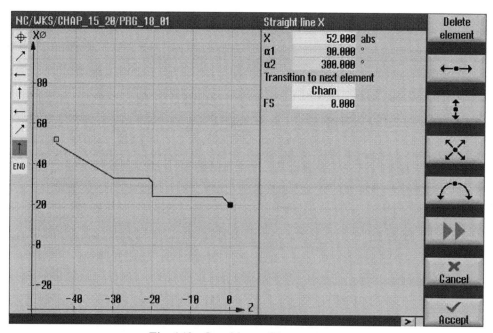

Fig. 140. Graphic profile generator

18.4 Practical exercise

18.4.1 How to identify a profile in a program

Open the program 'PRG_18_01' in the folder 'CHAP_15_20' This program cannot be executed. The following exercise allows you to become familiar with the four options for the definition of the profile offered by the cycle CYCLE62.
This cycle will be used in the following chapter as well.

Insert the cycle in the program where shown using the following data:

for option 1
select: Label
LAB1:   PROFILE1
LAB2:   END1

for option 2
select: Subprogram
PRG: PROFILE1

for option 3
select: Label in subprogram
PRG: EXT_PROF1
LAB1:   PROFILE1
LAB2:   END1

for option 4
select: Profile name
CON: EXTERNAL1

Compare your program to the one in the folder FINISHED_EXERCISES named EX_18_01 and shown on the next page.

```
; program not executable

G18 G54 G90
G0 X400 Z500
M8
SETMS(1)

T1 D1
G95 S1400 M4
G0 X62 Z2

; enter option 1
CYCLE62(,2,"PROFILE1","END1")

; enter option 2
CYCLE62("PROFILE1",0,,)

; enter option 3
CYCLE62("EXT_PROF1",3,"PROFILE1","END1")

; enter option 4
CYCLE62("EXTERNAL1",1,,)

G0 X200
G0 Z200

M30
```

19. CYCLE952: Roughing Cycle (2h)
(Theory: 1h, Practice: 1h)

19.1 Description
Important: the roughing cycle CYCLE952 removes the material according to the profile selected by the cycle CYCLE62.

The removal of excess material may be programmed in longitudinal, radial or parallel direction to the profile, either on outside or inside the workpiece. The meaning of the parameters is easily explained by animations and images in the dialog box of the cycle. For further information press the HELP button as described in paragraph 17.2.
Among the various options that the cycle offers, we will analyze the most efficient and logical way to use it, skipping over the settings which are useful in specific cases only.

19.2 Insertion procedure
Starting from the main menu shown in paragraph 17.3, proceed by pressing the buttons shown in the following table:

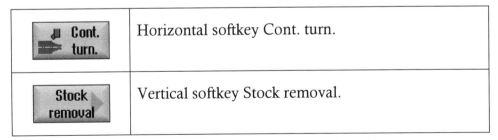

Fig. 141. CYCLE952: procedure for the insertion of the cycle

19.3 Parameters

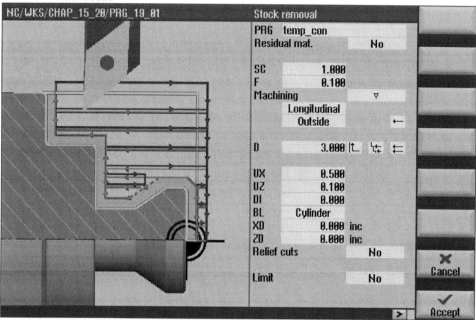

Fig. 142. CYCLE952: window for the insertion of the parameters

Parameter	Description
PRG	Name of the temporary program the cycle refers to in order to create and save the tool path. Various types of names can be used (e.g. temp_con, temporary contour).
Residual mat.	Possibility to save the updated profile of the blank part in order to carry out the next tooling operation on the remaining material. With the SELECT button, select the options: - YES - NO (recommended selection).
CONR	When pressing YES, enter the name with which you want to save the updated profile.

SC	Safety distance for the approach of the tool used on the X-axis with diametral value and on the Z-axis with real value (example for a value: 1 mm).
F	Feedrate used by the tool during the roughing phase. The feedrate programmed in the finished profile of the workpiece is not taken into account in this case.
Machining ▽ ▽▽▽	Type of tooling operation to carry out. Roughing of the workpiece (please select). Only finishing of the selected profile.
Longitudinal Face Parallel to the contour	For workpieces with proportions similar to a shaft. For workpieces with proportions similar to a flange. For workpieces already shaped as casting, molded parts or preworked parts.
Outside Inside RP	For external profiles. For internal profiles. If internal, it expresses the return value in X.
← →	Inversion of the cutting direction. The NC checks if this selection is compatible with the position of the cutting edge.
D	Pass depth (if it has a radial value in X).

⌐	Adapts the pass trajectories to the profile to be roughed and levels the shoulders, thereby guaranteeing the possibility to leave a constant machining allowance on the whole profile (recommended selection).
←	Exclusively performs straight trajectories leaving a stepped surface even in the presence of cones.
⌐	Adapts the pass direction to the trajectories of the profile to be roughed by executing them only when the angle between the cutting edge and the profile exceeds the value determined by the machine data.
↳	Refers the distribution of the passes to the edges of the profile while maintaining the pass depth as constant as possible (to be used when, due to problems with chip control, it is preferable not to execute passes of shallow depth).
↳	The pass always starts with the set cutting depth and then adapts to the edges of the workpiece (allows the pre-emptive calculation of the pass number necessary to carry out the roughing operation).
← ←	Constant cutting direction (recommended choice). Inclined cutting direction (prevents the craterization of the insert).
UX	Allowance to be left on the X-axes with radial value.
UZ	Allowance to be left on the Z-axes.
DI	Cutting length in millimeters after which the tool moves back in order to break the chip. If the value is zero, a continuous pass without chip breaking movements will be carried out.

BL	Description of the distribution of the material to be removed. - Cylinder: material quantity to be removed beyond the cylinder defined by the end points of the profile. - Allowance: total quantity of material to be removed for shaped workpieces. - Contour: specific definition of the profile of rough material to be removed.
XD	Further quantity of rough material in X (used for blank parts turning eccentrically).
ZD	Further quantity of rough material in Z (used for non-orthogonal faces on the Z-axis).
Relief cuts FR	With the SELECT button, select: - YES in order to execute the tooling operation of the parts on the shadow side of the profile. - NO in order to rough the profile without taking into consideration the parts on the shadow side. When selecting YES the cycle asks that a specific feedrate be entered for plunge cutting.
Limit XA XB ZA ZB	With the SELECT button, select: - YES in order to limit the material to be roughed in a certain area. - NO in order to remove all the material. When selecting YES the cycle asks that you enter the absolute coordinates limiting the area of the material to be removed.

Fig. 143. CYCLE952: list of cycle parameters

19.4 Practical exercise

19.4.1 External roughing and finishing of a workpiece

Open the program 'PRG_19_01' in the folder 'CHAP_15_20'
This program executes the roughing of the workpiece previously analyzed:
- in paragraph 11.5.1 the workpiece was roughed using the point to point programming of each single of the tool's working, disengagement and repositioning passes.
- now, the following program defines the reference profile using the cycle CYCLE62, roughs the workpiece using the cycle CYCLE952, to then proceed to finishing with another tool, executing the same profile as previously selected.

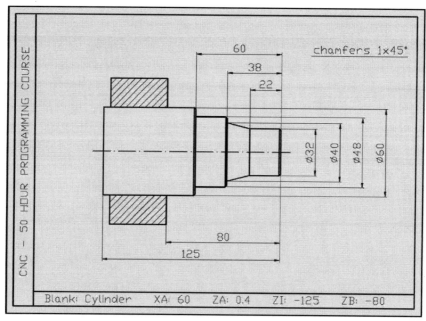

Fig. 144. External roughing of a workpiece with fixed cycle

After the start of the graphic simulation and having observed the tool movements, proceed to change the parameters of the roughing cycle on the basis of the description given in paragraph 19.3 and observe the variations in the execution of the tooling operation.

```
; blank part dimensions:
; XA = 60 bar diameter
; ZA = 0.4 machining allowance on front face
; ZI = -125 length of finished part
; ZB = -80 extension from jaws
N10 WORKPIECE(,,,"CYLINDER",192,0.4,-125,-80,60)

N20 G18 G54 G90
N30 G0 X400 Z500
N40 M8
N50 SETMS(1)

N60 T1 D1 ; ROUGHING TOOL
N70 G95 S1400 M4
N80 G0 X62 Z2 ; APPROACH OUTSIDE OF THE WORKPIECE

N90 CYCLE62(,2,"PROFILE1","END1")

N100
CYCLE952("temp_con",,"",2301311,0.1,0,0,3,0.1,0.1,0.5,0.1,0.1
,0,1,0,0,,,,,,2,2,,,0,1,,0,12,1100110,1,0)

N110 G0 X200
N120 G0 Z200

N130 T2 D1 ; FINISHING TOOL
N140 G95 S1800 M4
;EXECUTION OF THE FACING
N150 G1 X35 Z0
N160 G1 X-1

N170 G0 X28 Z2 ; STARTING POSITION OF THE PROFILE TO BE
FINISHED
```
N180 G1 X26 Z0 G42 ; ENABLING OF THE TOOL RADIUS COMPENSATION BEFORE THE PROFILE USED BY THE CYCLE

N190 PROFILE1:
```
N200 G1 X26 Z0 F0.1
N210 G1 X32 CHR=1 FRCM=0.04
N220 G1 Z-22
N230 G1 Z-38 ANG=166
```

```
N240 G1 X48 CHR=1
N250 G1 Z-60
N260 G1 X60 CHR=1
N270 G1 Z-62
N280 G1 X61
```
N290 END1:

```
N300 G0 X200 G40
N310 G0 Z200

N320 M30
```

19.4.2 Programming alternative with subprogram

When you open the program 'PRG_19_02' in the folder 'CHAP_15_20' you notice that the programming option entails the enclosure of the workpiece's finished profile within a subprogram.

The selection of the profile to be roughed is carried out by the cycle CYCLE62, though according to the instructions contained in option 2 in paragraph 18.3.

You will also see that in the finishing phase the profile of the workpiece has been replaced by the call of the subprogram.

This programming alternative has the advantage of shortening program length, but the disadvantage of hosting the workpiece profile outside the main program.

```
; blank part dimensions:
; XA = 60 bar diameter
; ZA = 0.4 machining allowance on front face
; ZI = -125 length of finished part
; ZB = -80 extension from jaws
N10 WORKPIECE(,,,"CYLINDER",192,0.4,-125,-80,60)

N20 G18 G54 G90
N30 G0 X400 Z500
N40 M8
N50 SETMS(1)

N60 T1 D1 ; ROUGHING TOOL
N70 G95 S1400 M4
N80 G0 X62 Z2 ; APPROACH OUTSIDE OF THE WORKPIECE
```

```
N90 CYCLE62("PROFILE1",0,,)

N100
CYCLE952("temp_con",,"",2101311,0.1,0,0,3,0.1,0.1,0.5,0.1,0.1
,0,1,0,0,,,,,,2,2,,,0,1,,0,12,1100110)

N110 G0 X200 Z200

N120 T2 D1 ; FINISHING TOOL
N130 G95 S1800 M4
;EXECUTION OF THE FACING
N140 G1 X35 Z0
N150 G1 X-1
N160 G0 X28 Z2 ; STARTING POSITION OF THE PROFILE TO BE
FINISHED
N170 G1 X26 Z0 G42
```

N180 PROFILE1 ; IDENTIFICATION OF THE PROFILE CONTAINED IN THE SUBPROGRAM

```
N190 G0 X200 Z200 G40

N200 M30
```

19.4.3 Internal roughing and finishing of a workpiece

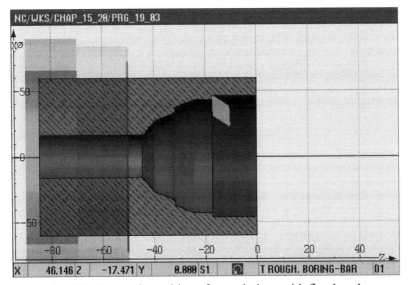

Fig. 145. Internal roughing of a workpiece with fixed cycle

Now open the program 'PRG_19_03' in the folder 'CHAP_15_20'

This program uses the roughing cycle to create the part previously analyzed in paragraph 15.5.1.
Start the graphic simulation and observe the tool movements. Proceed to change the cycle parameters on the basis of the description given in paragraph 19.3 and observe the variations in the execution of the tooling operation

```
; blank part dimensions:
; XA = 60 bar diameter
; ZA = 0.2 machining allowance on front face
; ZI = -85 length of finished part
; ZB = -50 extension from jaws
N10 WORKPIECE(,,,"CYLINDER",0,0.2,-85,-50,60)

N20 G18 G54 G90
N30 G0 X400 Z500
N40 M8
N50 SETMS(1)

N60 LIMS=3000 ; LIMITATION TO 3000 REV/MIN

N70 T1 D1 ; TURNING TOOL FOR EXTERNAL PARTS
N80 G96 S100 M4 ; ENABLING OF CONSTANT CUTTING SPEED
N90 G0 X62 Z0 ; APPROACH
N100 G1 X-1.6 F0.18 ; FACING
N110 G0 X200 Z200 ; DISENGAGEMENT

N120 T11 D1 ; RIGHT AXIAL DRILL DIAMETER 16 MM
N130 G95 S1100 M3 ; ENABLING OF FIXED NUMBER OF REVOLUTIONS
N140 G0 X0 Z2 ; APPROACH
N150 G1 Z-30 F0.12 ; FIRST DRILLING PASS
N160 G4 S2 ; DWELL TIME OF 2 SPINDLE REVOLUTIONS
N170 G0 Z5 ; RAPID EXIT FOR CHIP REMOVAL
N180 G1 Z-29 F2 ; ENTERING WITH HIGH FEEDRATE
N190 G1 Z-60 F0.12 ; SECOND PASS UP TO Z-60
N200 G4 S2
N210 G0 Z5
N220 G1 Z-59 F2
N230 G1 Z-90 F0.12
N240 G0 Z200 ; DISENGAGEMENT IN Z

N250 T12 D1 ; BORING BAR FOR INTERNAL TURNING
N260 G96 S120 M4 ; ENABLING OF CONSTANT CUTTING SPEED
```

```
N270 G0 X16 Z2

N280 CYCLE62(,2,"INTERNAL1","END1")

N290
CYCLE952("temp_con",,"",2102311,0.1,0,0,2,0.1,0.1,0.5,0.1,0.1
,0,1,0,0,,,,,,2,2,,,0,1,,0,12,1100110)

N300 G0 Z5
N310 G0 X48

;BEGINNING OF FINISHING WITH SAME TOOL
N320 G96 S150 M4 ; ENABLING OF CONSTANT CUTTING SPEED
N330 G1 X50 Z0 G41 F0.1 ; ENABLING OF THE TOOL RADIUS
COMPENSATION OUTSIDE OF THE DEFINITION OF THE PROFILE

N340 INTERNAL1:
N350 G1 X50 Z0
N360 G1 Z-1 ANG=225
N370 G1 Z-14
N380 G3 X38 Z-32 CR=58
N390 G1 X34 RND=2
N400 G1 ANG=201
N410 G1 X16 Z-45 ANG=230 RND=2
N420 G1 Z-48
N430 G1 X15
N440 END1:

N450 G0 Z5 G40
N460 G0 X200 Z200
N470 M30
```

19.4.4 Use of the tool radius compensation

The roughing cycle carries out the execution of the passes by automatically considering the dimension of the tool radius. **The cycle does not allow for the programming of the functions G41 and G42 in the selected profile, otherwise the following alarm is given:**

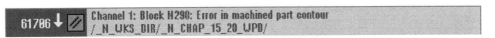

The program 'PRG_19_04' contained in the folder 'CHAP_15_20' uses CYCLE952 to execute both the roughing and the finishing of the shape. This is entered after the end of the program following the function M30. This solution helps avoid programming mistakes in relation to the activation of the tool radius compensation by delegation of the correction of the tool path to the cycle.

```
N10  WORKPIECE(,,,"CYLINDER",192,0,-125,-80,60)
N20  G18 G54 G90
N30  G0 X400 Z500
N40  SETMS(1)

N50  T1 D1 G95 S1400 M4 ; ROUGHING TOOL
N60  G0 X62 Z2 M8
N70  CYCLE62(,2,"PROFILE1","END1")

N80
CYCLE952("temp_con",,"",2301311,0.1,0,0,3,0.1,0.1,0.5,0.1,0.1
,0,1,0,0,,,,,,2,2,,,0,1,,0,12,1100110,1,0)
N90  G0 X200 Z200

N100 T2 D1 G95 S1800 M4 ; FINISHING TOOL
N110 G0 X62 Z2
N120 CYCLE62(,2,"PROFILE1","END1")

N130
CYCLE952("temp_con",,"",2301321,0.1,0,0,3,0.1,0.1,0.5,0.1,0.1
,0,1,0,0,,,,,,2,2,,,0,1,,0,12,1100110,1,0)
N140 G0 X200 Z200

N150 M30

N160 PROFILE1:
N170 G1 X26 Z0 F0.1
N180 G1 X32 CHR=1 FRCM=0.04
N190 G1 Z-22
N200 G1 Z-38 ANG=166
N210 G1 X48 CHR=1
N220 G1 Z-60
N230 G1 X60 CHR=1
N240 G1 Z-62
N250 G1 X61
N260 END1:
```

20. CYCLE99: Threading Cycle (1h)
(Theory: 0.5h, Practice: 0.5h)

20.1 Description

The cycle executes a complete threading in a single block.
It is able to suggest the dimensions of the threading to be created, to distribute the passes while keeping the chip section or the pass depth constant, to cut on one side of the tool or to use both cutting edges.
It is always more practical and faster to use the cycle instead of programming the threading with ISO functions as seen in paragraph 11.3.

20.2 Insertion procedure

Starting from the main menu shown in paragraph 17.3, proceed by pressing the buttons shown in the following table:

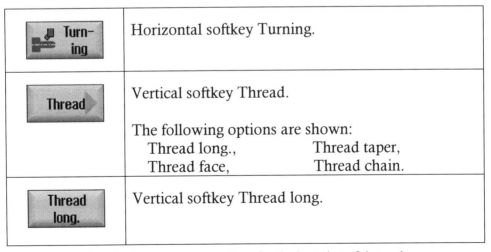

Fig. 146. CYCLE99: procedure for the insertion of the cycle

20.3 Parameters

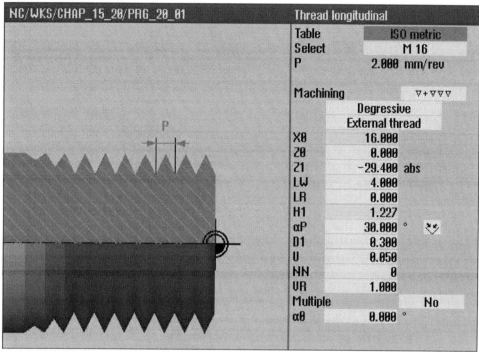

Fig. 147. CYCLE99: window for the insertion of the parameters

Parameter	Description
Table	Type of table from which to automatically extract the characteristic data of the threading to insert into the respective parameters. - Without: the help of the table is not necessary, - ISO metric, - Whitworth BSW (British Standard Whitworth), - Whitworth BSP (British Standard Pipe), - UNC (Unified Coarse Thread).
Selection	Enter the thread you want to form.

P	Thread lead. Automatically proposed if you use the TABLE data.
Machining	Type of tooling operation to carry out.
▽	Execution of the threading without elimination of the machining allowance.
▽▽▽	Execution of the finishing operation only.
▽+▽▽▽	Complete execution of the threading.
Linear	Constant pass depth.
Digressive	Constant chip section.
External thread	External threading.
Internal thread	Internal threading.
X0	Starting diameter of the thread (less than the diameter of the hole for internal threads).
Z0	Starting value of the thread expressed in absolute coordinates on the Z-axis.
Z1	Arrival value of the thread expressed in absolute coordinates (recommended choice) on the Z-axis and **referred to the tool zero point**. With the SELECT button you can set the arrival point of the thread as an incremental value.

LW	Length of the approach path at the starting point of the thread on the Z-axis (recommended choice). This value corresponds to approx. two times the lead. With the SELECT button you can change the meaning of the parameter.
LW2	Length of the entry path into the thread.
LW2=LR	Length of the entry path into the thread equals the length of the exit path expressed in parameter LR.
LR	Length of the exit path out of the thread. If the value is zero the exit is carried out at 90°.
H1	Radial depth of the thread.
DP	Inclination of the profile expressed as length.
αP	Angle of the profile expressed in degrees (recommended). This value equals 30 for a metric threading and 27.5 for a Whitworth threading.
	Select the option by pressing the SELECT button.
⬇	The passes are executed by alternating the cutting side of the tool.
⬇	The passes are executed following the inclination direction of the thread.
D1	Radial depth of the first pass.

U	Radial depth of the last pass.
NN	Number of dry cuts.
VR	Radial separation value in X.
Multiple leads	Number of threads to create.
α0	Thread angle.

20.4 Practical exercise

20.4.1 Programming example
Open the program 'PRG_20_01' in the folder 'CHAP_15_20'
This program uses the threading cycle to create the part previously analyzed in paragraph 11.5.3.
Start the graphic simulation and observe the tool movements. Proceed to change the cycle parameters on the basis of the description given in paragraph 20.3 and observe the variations in the execution of the tooling operation.

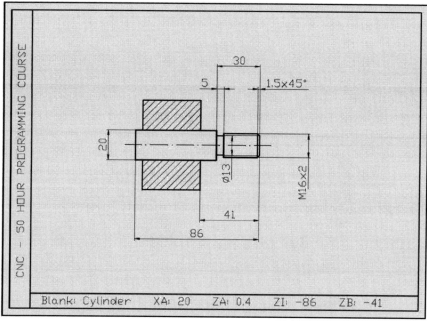

Fig. 148. Example of the programming of a threaded workpiece

```
; blank part dimensions:
; XA = 20 bar diameter
; ZA = 0.4 machining allowance on front face
; ZI = -86 length of finished part
; ZB = -41 extension from jaws
WORKPIECE(,,,"CYLINDER",0,0.4,-86,-41,20)

G18 G54 G90
G0 X400 Z500
M8
```

```
SETMS(1)

LIMS=3000
T1 D1 ; EXTERNAL TURNING
G96 S100 M4
G0 X22 Z0
G1 X-1.6 F0.18
G0 X12.8 Z0.5
G1 Z0
G1 X15.8 Z-1.5
G1 Z-30
G1 X22
G0 X200
G0 Z200

T3 D1 ; GROOVING TOOL WIDTH 3MM
G95 S800 M4
G0 Z-30
G0 X22
G1 X13 F0.1
G4 S2
G0 X22
G0 Z=IC(2)
G1 X13
G4 S2
G0 X22
G0 X200
G0 Z200

T4 D1 ; TOOL FOR EXTERNAL THREADS
G95 S600 M3
G0 Z4
G0 X20

CYCLE99(0,16,-
29.4,,4,0,1.2268,0.05,30,0,5,0,2,1310203,4,2,0.3,0.5,0,0,1,0,
0.708293,1,,"ISO_METRIC","M16",1102,0)

G0 X200 ; DISENGAGEMENT
G0 Z200
M30
```

21. Second Test (2h)
(Practice: 2h)

21.1 Introduction to the test

The test consists in the execution of the program creating the part shown in figure 152. Take the following steps:
- Load the tools file which you find in folder 01_EXERCISES named EMPTY_TOOL_LIST. This file deletes all the existing tools by overwriting them only with the roughing tool defined therein. In order to load this file, follow the procedure laid down in paragraph 3.3.
- Now create the necessary tools for the execution of this program following the procedure specified in paragraph 7.5.1. Below is a list of the necessary tools, their position in the turret, the offset data in X and Z and the data for the definition of their graphic aspect.

1	ROUGHING TOOL	1	1	88.000	40.000	0.800 ←	93.0	55	11.0
2									
3	OD GROOVING W.3MM	1	1	98.000	40.000	0.100	3.000		10.0
4									
5	CENTER DRILL D.6	1	1	100.000	24.000	6.000	118.0		
6									
7									
8									
9									
10									
11	AX. DRILL D.16	1	1	100.000	120.000	16.000	118.0		
12	ROUGH. BORING-BAR	1	1	86.000	92.000	0.400 ←	93.0	55	8.0
13	FINISH. BORING-BAR	1	1	84.000	88.000	0.200 ←	93.0	55	8.0
14	ID GROOV. W.3MM	1	1	92.000	75.000	0.100	3.000		8.0
15	ID THREADING	1	1	88.000	95.000	0.200			
16	AX. DRILL D.12	1	1	100.000	72.000	12.000	118.0		

Fig. 149. List of tools to be created and used in the test program

- In position 14, it is necessary to create a tool for internal grooves with a width of 3 millimeters and with its zero point set on the left. Press the 'Next' arrow shown in the following figure to get to the point where you can choose the correct tool.

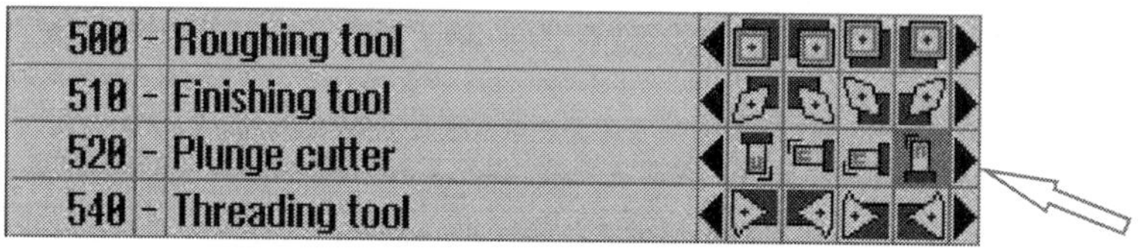

Fig. 150. Choice of the position of the cutting edge

- In the folder 01_EXERCISES, create an empty main program and name it TEST_20_01.
- Structure the program as the ones we've seen so far.
- Enter the comments with the dimensions of the blank part at the beginning of the program. If you want to copy blocks from already existing programs see paragraph 14.4.
- Define the dimensions of the blank part according to the procedure laid down in paragraph 12.6.2.
- Enter the blocks that activate the initial settings and the home position:
```
G18 G54 G90
G0 X400 Z500
M8
SETMS(1)
```
- Proceed to the programming of the tooling operations following the logical sequence described in paragraph 5.2.

21.2 Tooling operations and cutting parameters

Tooling sequence	Tool	Operation	Cutting speed (m/min)	Feedrate (mm/rev)
1st	T1 D1	External profile	100	0.18
2nd	T3 D1	External grooves	78	0.1
3rd	T5 D1	Center drilling	80	0.07
4th	T16 D1	Hole D12	60	0.1
5th	T12 D1	Int. roughing	70	-
6th	T13 D1	Int. finishing	90	-
7th	T14 D1	Int. groove	60	-
8th	T15 D1	Int. threading	60	-

Fig. 151. Sequence of tooling operations and cutting parameters to use for the test

- Execute the 1st operation in a single pass.
- Execute center drilling up to Z-5.
- Execute the hole in four passes, programming the exit of the drill at Z2 in order to remove the chip.
- Use the roughing cycle for the execution of the 5th tooling operation.
- Use constant cutting speed for the execution of the 5th and the 6th tooling operation.

21.3 Drawing of the part to be created

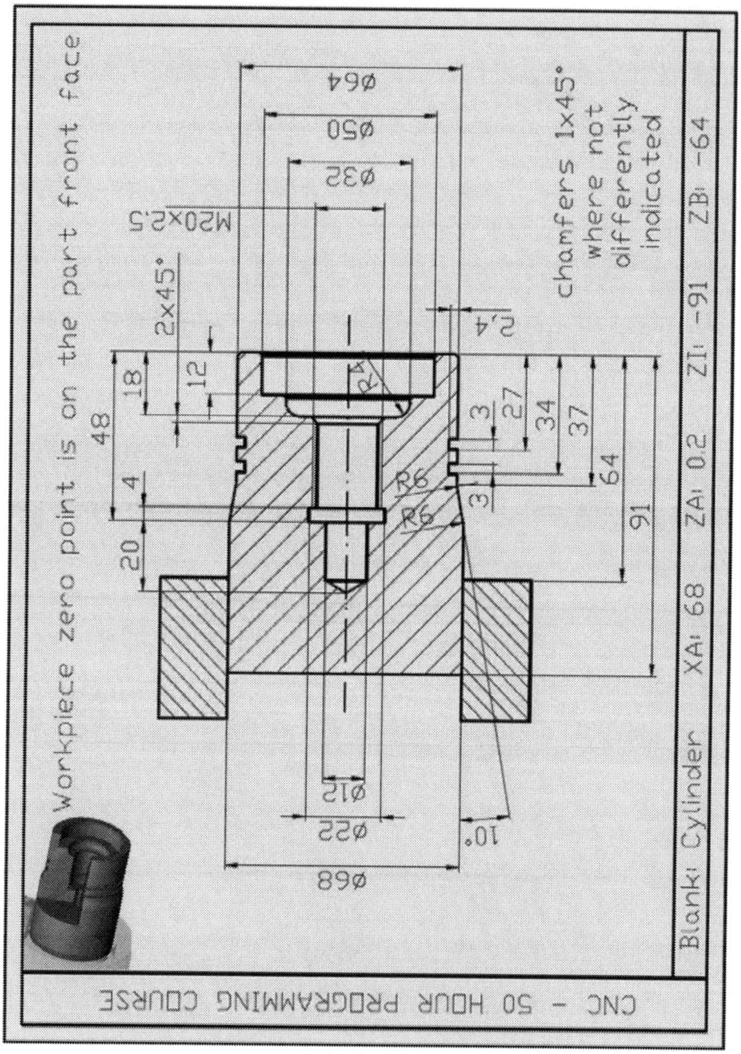

Fig. 152. Drawing of the part to be created

21.4 Program correction and reloading of tool list
Compare your program to the one in the folder FINISHED_EXERCISES named TEST_20_01.
Before reading the next chapter, reload the complete tool list contained in the folder 01_EXERCISES.

22. CYCLE930: Cycle for Grooves (1h)
(Theory: 0.5h, Practice: 0.5h)

22.1 Description
This cycle executes radial or frontal, internal or external grooves with straight or with inclined walls on cylindrical or conical diameters, and creates chamfers or radii on external and internal edges. On the basis of the cycle's settings you can first carry out the roughing and then, after the tool change, the finishing of the groove. You can furthermore execute the automatic chip breaking during the first pass and create a series of identical grooves.

22.2 Insertion procedure
Starting from the main menu shown in paragraph 17.3, proceed by pressing the buttons shown in the following table:

 Turning	Horizontal softkey Turning.
 Groove	Vertical softkey Groove. The following options are shown: - Groove with straight walls without chamfers and/or radii on the edges. - Groove with possible straight walls with chamfers and/or radii on the edges. - Groove on conical diameter.
 ⊔	Choose the second option.

Fig. 153. CYCLE930: procedure for the insertion of the cycle

22.3 Parameters

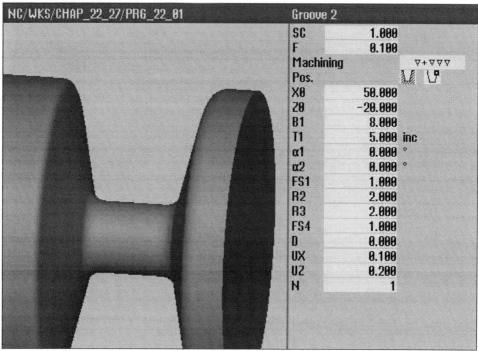

Fig. 154. CYCLE930: window for the insertion of the parameters

Parameter	Description
SC	Approaching distance. Has diametral value when referring to the X-axes.
F	Feedrate.
Machining	Type of tooling operation to carry out.
▽	Execution of the roughing of the groove without elimination of the machining allowance.
▽▽▽	Execution of the finishing operation only.
▽+▽▽▽	Complete execution of the groove.

Position	**Orientation of the groove.** Select the option by pressing the SELECT button.
⊔	External radial groove.
⊏	Anterior frontal groove.
⊓	Internal radial groove.
⊐	Posterior frontal groove.
	Position of the starting point of the groove.
⌐⊔	Starting point on external diameter on the right.
⌐⊔	Starting point on internal diameter on the right.
⊔⌐	Starting point on internal diameter on the left.
⊔⌐	Starting point on external diameter on the left.
X0	Diameter on which the starting point lies.
Z0	Position in Z of the starting point.
B1	Width of the groove.
T1	Depth of the groove. With the SELECT button, choose if you want to use an absolute value or an incremental value with respect to the starting point (on the X-axes, the incremental value as radial meaning).
α1	Inclination angle of the left wall.

α2	Inclination angle of the right wall.
FS1 / R1	Select the option by pressing the SELECT button. - FS1: chamfer on the external right edge. - R1: radius on the external right edge.
FS2 / R2	Select the option by pressing the SELECT button. - FS2: chamfer on the internal right edge. - R2: radius on the internal right edge.
FS3 / R3	Select the option by pressing the SELECT button. - FS3: chamfer on the internal left edge. - R3: radius on the internal left edge.
FS4 / R4	Select the option by pressing the SELECT button. - FS4: chamfer on the external left edge. - R4: radius on the external left edge.
D	Length of the pass after which to execute a chip break during the first entry into the groove. For continuous cutting enter 0.
UX	Machining allowance with radial value on X-axis.
UZ	Machining allowance on Z-axis.
N	Number of grooves to create in a series.
DP	If N is higher than 1, DP expresses the lead of the grooves.

22.4 Practical exercise

22.4.1 Programming example
Open the program 'PRG_22_01' in the folder 'CHAP_22_27'
This program uses the cycle for grooves to create the part shown in the following figure.
Start the graphic simulation and observe the tool movements. Proceed to change the cycle parameters on the basis of the description given in paragraph 22.3 and observe the variations in the execution of the tooling operation.

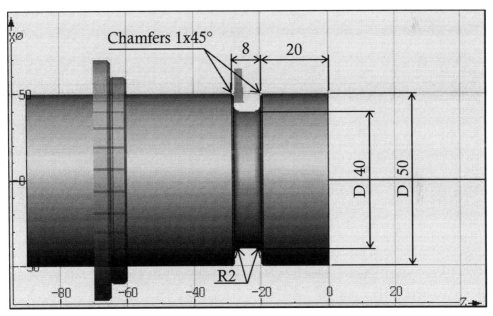

Fig. 155. Programming example with the use of the groove cycle

```
; blank part dimensions:
; XA = 50 bar diameter
; ZA = 0 machining allowance on front face
; ZI = -90 length of finished part
; ZB = -60 extension from jaws
N10 WORKPIECE(,,,"CYLINDER",0,0,-90,-60,50)

N20 G18 G54 G90
N30 G0 X400 Z500
N40 M8
N50 SETMS(1)
```

```
N60 LIMS=3000

N70 T3 D1 ; GROOVING TOOL WIDTH 3 MM
N80 G96 S80 M4
N90 G0 Z-23
N100 G0 X52


N110 CYCLE930(50,-
20,8,8,5,,0,0,0,1,2,2,1,0.2,0,1,10530,,1,30,0.1,1,0.1,0.2,2,1
001110)

N120 G0 X200
N130 G0 Z200

N140 M30
```

23. CYCLE82: Drilling Cycle (1h)
(Theory: 0.5h, Practice: 0.5h)

23.1 Description
The drilling cycle creates longitudinal or radial holes, allowing to specify the approaching distance, the drilling depth, the dwell time at the bottom of the hole and the disengagement distance at the end of the execution by means of its parameters.

The cycle does not have a specific parameter for the setting of the feedrate; it is therefore necessary to program it before the cycle activation.

23.2 Insertion procedure
Starting from the main menu shown in paragraph 17.3, proceed by pressing the buttons shown in the following table:

 Drilling	Horizontal softkey Drilling.
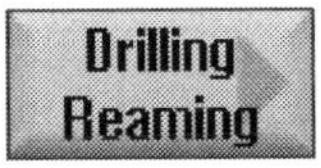 Drilling Reaming	Vertical softkey Drilling Reaming The reaming cycle is similar to the drilling cycle, but it has the additional option to set a working feedrate to get into the hole and a faster rate to exit from the hole.
 Drilling	Select Drilling.

Fig. 156. CYCLE82: procedure for the insertion of the cycle

23.3 Parameters

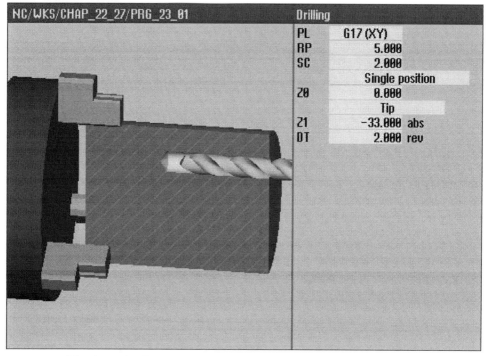

Fig. 157. CYCLE82: window for the insertion of the parameters

Parameter	Description
PL	Definition of the plane perpendicular to the translation axis of the drill. - G17 for holes along Z. - G18 for holes along Y (not applicable to this machine). - G19 for holes along X.
RP	Absolute value for the disengagement after the tooling operation.
SC	Incremental approaching distance referring to the starting point Z0.

Single position	Setting of the cycle for the execution of a single hole.
Position pattern (MCALL)	Modal activation of the drilling cycle for the execution of multiple holes (see par. 23.4.3).
Z0	Absolute starting coordinate for the hole.
Tip	Parameter Z1 refers to the tip of the drill.
Shank	Parameter Z1 refers to the end of the cylindrical part of the drill.
Z1	Arrival point of the drilling. Select the option by pressing the SELECT button.
abs	The value of Z1 has absolute meaning referring to the part zero point.
inc	The value of Z1 has incremental meaning with respect to the starting point expressed by the parameter 'Z0'.
DT	Dwell time at the bottom of the hole before the disengagement of the drill. Select the option by pressing the SELECT button.
rev	The dwell time is expressed in spindle revolutions.
s	The dwell time is expressed in seconds.

23.4 Practical exercise

23.4.1 Example of the programming of an axial hole
Open the program 'PRG_23_01' in the folder 'CHAP_22_27'
This program uses the drilling cycle to create the part shown in the following figure.
Start the graphic simulation and observe the tool movements. Proceed to change the cycle parameters on the basis of the description given in paragraph 23.3 and observe the variations in the execution of the tooling operation.

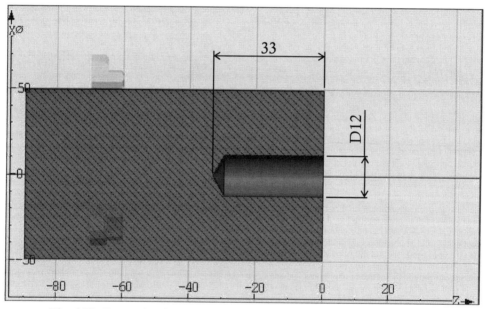

Fig. 158. Example of the programming of an axially drilled workpiece

```
; blank part dimensions:
; XA = 50 bar diameter
; ZA = 0 machining allowance on front face
; ZI = -90 length of finished part
; ZB = -60 extension from jaws
N10 WORKPIECE(,,,"CYLINDER",0,0,-90,-60,50)

N20 G18 G54 G90
N30 G0 X400 Z500
N40 M8
N50 SETMS(1)
```

```
N60 T16 D1 ; FIXED AXIAL DRILL D12
N70 G95 S1200 M3 F0.1
N80 G0 Z20
N90 G0 X0

N100 CYCLE82(5,0,2,-33,,2,0,1,22)

N110 G0 Z200
N120 G0 X200
N130 M30
```

23.4.2 Example of the programming of a radial hole

Open the program 'PRG_23_02' in the folder 'CHAP_22_27' and proceed as in the previous paragraph.

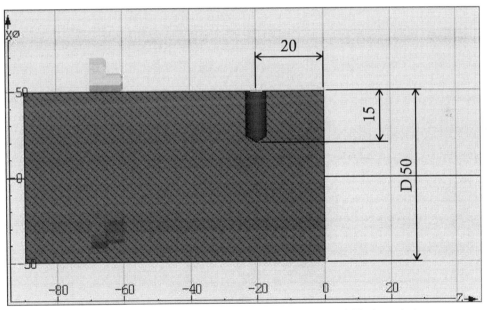

Fig. 159. Example of the programming of a radially drilled workpiece

```
; blank part dimensions:
; XA = 50 bar diameter
; ZA = 0 machining allowance on front face
; ZI = -90 length of finished part
; ZB = -60 extension from jaws
N10 WORKPIECE(,,,"CYLINDER",0,0,-90,-60,50)

N20 G18 G54 G90
```

```
N30  G0 X400 Z500 M8
N40  SETMS(3)
N50  SPOS[1]=0

N60  T8 D1 ; DRIVEN RADIAL DRILL D6
N70  G95 S1200 M3 F0.1
N90  G0 Z-20
N100 G0 X55
N110 CYCLE82(55,50,2,20,,2,0,3,22)
N120 G18 ; RETURN TO THE TURNING PLANE X-Z

N130 G0 X300
N140 G0 Z300
N150 M30
```

23.4.3 Modal activation of the cycle with MCALL

Open the program 'PRG_23_03' in the folder 'CHAP_22_27' and start the graphic simulation. The MCALL function allows for the modal activation of the drilling cycle, which will be repeated for all the positions programmed subsequently; it is disabled with the command MCALL, followed by nothing.

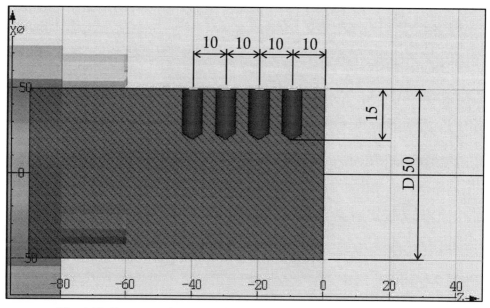

Fig. 160. Programming example with modal activation of the cycle

```
; blank part dimensions:
; XA = 50 bar diameter
; ZA = 0 machining allowance on front face
; ZI = -90 length of finished part
; ZB = -60 extension from jaws
N10 WORKPIECE(,,,"CYLINDER",192,0,-90,-60,50)

N20 G18 G54 G90
N30 G0 X400
N40 G0 Z500
N50 M8
N60 SETMS(3)
N70 SPOS[1]=0 ; ANGULAR ORIENTATION OF THE WORKPIECE

N80 T8 D1 ; DRIVEN RADIAL DRILL D6
N90 G95 S1200 M3 F0.1
N100 G0 Z-10
N110 G0 X55

N120 G19 ; DEFINITION OF THE WORKING PLANE Y-Z
N130 MCALL CYCLE82(55,50,2,20,,2,0,3,22)
N140 G0 Z-10
N150 G0 Z-20
N160 G0 Z-30
N170 G0 Z-40
N180 MCALL
N190 G18 ; RETURN TO THE TURNING PLANE X-Z

N200 G0 X300
N210 G0 Z300
N220 M30
```

The spindle of the driven tool is set as master spindle by the function SETMS in block N60.

The angular orientation of the workpiece is programmed in block N70 by means of the SPOS function.

The feedrate is programmed before the activation of the cycle in block N90 together with the number of revolutions of the tool.

Before the modal activation of the cycle it is necessary to define the working plane on which the holes to be drilled are located.

24. CYCLE83: Deep Hole Drilling Cycle (1h)
(Theory: 0.5h, Practice: 0.5h)

24.1 Description
This cycle is useful for the creation of particularly long holes which need to be executed with multiple passes.
The deep hole drilling cycle is very similar to the drilling cycle analyzed in the previous chapter, but has the additional option to program the automatic chip breaking or its removal from the hole. As in the drilling cycle, also this cycle does not have a specific parameter for the setting of the feedrate, which must be programmed before its activation. In the cycle there are the parameters for the progressive reduction of the feedrate with increasing pass depth.

24.2 Insertion procedure
Starting from the main menu shown in paragraph 17.3, proceed by pressing the buttons shown in the following table:

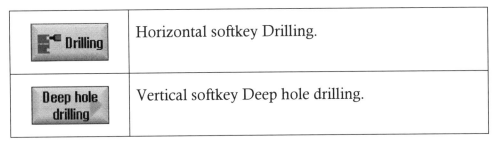

Drilling	Horizontal softkey Drilling.
Deep hole drilling	Vertical softkey Deep hole drilling.

Fig. 161. CYCLE83: procedure for the insertion of the cycle

24.3 Parameters

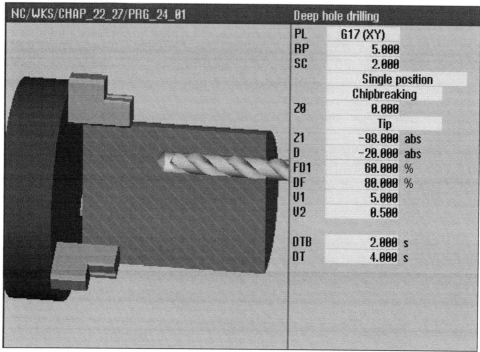

Fig. 162. CYCLE83: window for the insertion of the parameters

Parameter	Description
PL	Definition of the plane perpendicular to the translation axis of the drill. - G17 for holes along Z. - G18 for holes along Y (not applicable to this machine). - G19 for holes along X.
RP	Absolute value for the disengagement after the tooling operation.
SC	Incremental approaching distance referring to the starting point Z0.

Single position	Setting of the cycle for the execution of a single hole.
Position pattern (MCALL)	Modal activation of the drilling cycle for the execution of multiple holes.
Chip breaking	Select the option by pressing the SELECT button. Activation of the chip breaking cycle (without removal). The specific parameters are shown in paragraph 24.4.
Chip removal	Activation of the chip removal cycle. The specific parameters are shown in paragraph 24.5.
Z0	Absolute starting coordinate for the hole.
Tip	Select the option by pressing the SELECT button. Parameter Z1 refers to the tip of the drill.
Shank	Parameter Z1 refers to the end of the cylindrical part of the drill.
Z1	Arrival point of the drilling.
abs	Select the option by pressing the SELECT button. The value of Z1 has absolute meaning referring to the part zero point.
inc	The value of Z1 has incremental meaning with respect to the starting point expressed by the parameter 'Z0'.

D	Depth of the first pass.

Select the option by pressing the SELECT button. |
| abs | The value of D has absolute meaning referring to the part zero point. |
| inc | The value of D has incremental meaning with respect to the starting point expressed by the parameter 'Z0'. |
| FD1 | Reduction of the feedrate (the one programmed before the cycle) for the execution of the first pass expressed at parameter D.
This value is a percentage. |
| DF | Progressive reduction in percent of the pass length. The calculation always refers to the previous pass.

If DF is 100, this means that you do not want to decrement the pass lengths and that you want to keep them identical to the first one expressed at parameter D. In this case, the following parameter V1 is not displayed. |
| V1 | Minimum pass length beyond which the decremental calculation is interrupted. |

24.4 Parameters relative to the chip breaking

Parameter	Description
V2	Incremental retraction distance to execute at the end of every pass to guarantee the break of the chip.
DTB	Dwell time at the end of every pass before the incremental retraction.
	Select the option by pressing the SELECT button.
rev	The dwell time is expressed in spindle revolutions.
s	The dwell time is expressed in seconds.
DT	Specific dwell time at the end of the last pass.
	Select the option by pressing the SELECT button.
rev	The dwell time is expressed in spindle revolutions.
s	The dwell time is expressed in seconds.

24.5 Parameters concerning the chip removal

Parameter	Description
Lead distance	The distance between the bottom of the hole and the end position of the tip before the start of every pass. Select the option by pressing the SELECT button.
automatically	The cycle automatically manages this value by increasing it with higher hole depths.
manual V3	The distance is fixed and is manually set under parameter V3. Enter the value (e.g. 1 mm).
DTB rev s	Dwell time at the end of every pass before the disengagement of the drill. Select the option by pressing the SELECT button. The dwell time is expressed in spindle revolutions. The dwell time is expressed in seconds.
DT rev s	Specific dwell time at the end of the last pass. Select the option by pressing the SELECT button. The dwell time is expressed in spindle revolutions. The dwell time is expressed in seconds.
DTS rev s	Dwell time outside of the hole for the chip removal. Select the option by pressing the SELECT button. The dwell time is expressed in spindle revolutions. The dwell time is expressed in seconds.

24.6 Practical exercise

24.6.1 Execution of a hole with chip break
Open the program 'PRG_24_01' in the folder 'CHAP_22_27'
This program uses the drilling cycle to create the part shown in the following figure.
Start the graphic simulation and observe the tool movements. Proceed to change the cycle parameters on the basis of the description given in paragraph 24.3 and observe the variations in the execution of the tooling operation.

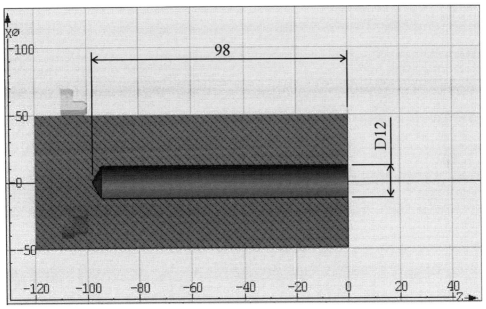

Fig. 163. Example of the programming of a drilled workpiece

```
; blank part dimensions:
; XA = 50 bar diameter
; ZA = 0 machining allowance on front face
; ZI = -120 length of finished part
; ZB = -100 extension from jaws
N10 WORKPIECE(,,,"CYLINDER",0,0,-120,-100,50)

N20 G18 G54 G90
N30 G0 X400 Z400
N40 M8
N50 SETMS(1)
```

```
N60 T16 D1 ; FIXED AXIAL DRILL D12
N70 G95 S1000 M3 F0.1
N80 G0 Z20
N90 G0 X0

N100 CYCLE83(5,0,2,-98,,-
20,,80,2,6,60,0,0,5,0.5,4,1,0,1,11221112)

N110 G0 Z200
N120 G0 X200
N130 M30
```

24.6.2 Execution of a hole with chip removal

Open the program 'PRG_24_02' in the folder 'CHAP_22_27' and proceed as in the previous paragraph.

This program creates the hole shown in figure 163 and performs the chip removal at every pass.

```
; blank part dimensions:
; XA = 50 bar diameter
; ZA = 0 machining allowance on front face
; ZI = -120 length of finished part
; ZB = -100 extension from jaws
N10 WORKPIECE(,,,"CYLINDER",0,0,-120,-100,50)

N20 G18 G54 G90
N30 G0 X400 Z400
N40 M8
N50 SETMS(1)

N60 T16 D1 ; FIXED AXIAL DRILL D12
N70 G95 S1000 M3 F0.1
N80 G0 Z20
N90 G0 X0

N100 CYCLE83(5,0,2,-98,,-
20,,80,0.5,0.5,60,1,0,5,0.5,2,1.6,0,1,11222112)

N110 G0 Z200
N120 G0 X200
N130 M30
```

25. CYCLE84/840: Tapping Cycle (1h)
(Theory: 0.5h, Practice: 0.5h)

25.1 Description
This cycle allows for the execution of rigid and compensated tappings by controlling the acceleration when entering the hole, the deceleration and the stop at arrival point, the inversion of rotation and the acceleration until exit from the hole for completion of the tapping process.

In order to execute a rigid tapping it is necessary that the machine be able to perfectly synchronize the spindle rotation with the tool translation. This feature allows the tapping tool to be mounted on a rigid support as if it were a drill bit; alternatively, one can choose compensated tapping, where the tool is mounted on a mobile support permitting the tool to move along the rotating axis, thereby compensating the position error of the slide.

25.2 Insertion procedure
Starting from the main menu shown in paragraph 17.3, proceed by pressing the buttons shown in the following table:

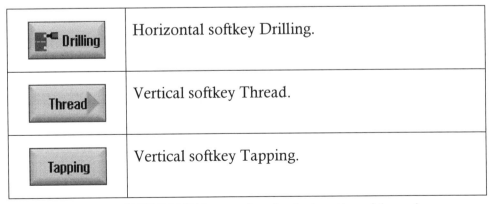

Fig. 164. CYCLE84/840: procedure for the insertion of the cycle

232

25.3 Parameters

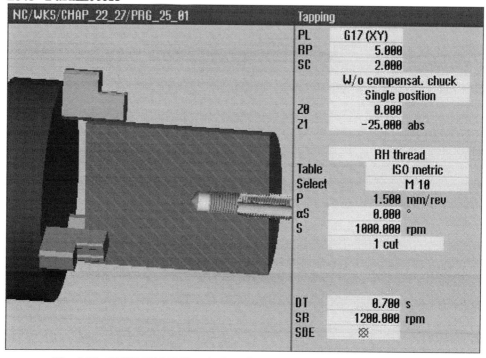

Fig. 165. CYCLE84/840: window for the insertion of the parameters

Parameter	Description
PL	Definition of the plane perpendicular to the translation axis of the tap. - G17 for tapping along Z. - G18 for tapping along Y (not applicable to this machine). - G19 for tapping along X.
RP	Absolute value for the disengagement after the tooling operation.
SC	Incremental approaching distance referring to the starting point Z0.

With compensat. holder	**CYCLE840:** the tool is mounted on a compensating holder.
W/o compensat. holder	**CYCLE84:** the tool is mounted on a rigid support like a drill (for rigid tapping).
Single position	Setting of the cycle for the execution of a single tapping.
Position pattern (MCALL)	Modal activation of the cycle for the execution of a series of tappings.
Z0	Absolute starting coordinate for the tapping.
Z1	Arrival point of the tapping. Select the option by pressing the SELECT button.
abs	The value of Z1 has absolute meaning referring to the part zero point.
inc	The value of Z1 has incremental meaning with respect to the starting point expressed by the parameter 'Z0'.
RH thread	The thread is a right hand thread.
LH thread	The thread is a left hand thread.

Table	Type of table from which to automatically extract the characteristic data of the thread to insert into the respective parameters. - Without: the help of the table is not necessary, - ISO metric, - Whitworth BSW (British Standard Whitworth), - Whitworth BSP (British Standard Pipe), - UNC (Unified Coarse Thread).
Selection	Enter the thread you want to form.
P	Thread lead. Automatically proposed if you use the TABLE data.
αS	Stagger from the angular starting position.
S	Fixed number of spindle revolution with which to execute the tapping
1 cut	The complete tapping is executed in one pass.
Chip breaking	At programmable intervals, the cycle inverts the rotation of the spindle to break the chip. The specific parameters are shown in paragraph 25.4.
Chip removal	At programmable intervals, the cycle inverts the rotation of the spindle and returns to the starting position (parameter SC) in order to remove the chip. The specific parameters are shown in paragraph 25.5.

DT	Dwell time before the inversion of the spindle rotation.
rev s	Select the option by pressing the SELECT button. The dwell time is expressed in spindle revolutions. The dwell time is expressed in seconds.
SR	Number of revolutions with which to exit from the hole with the tapping tool.
SDE	Rotation direction at cycle end. Allows for the setting of the spindle rotation direction preparing it for the next tooling operation. Select the option by pressing the SELECT button.
⊗	Spindle stop.
↻	Clockwise rotation (M3).
↺	Counterclockwise rotation (M4).

25.4 Parameters concerning the chip break

Parameter	Description
D	Length of pass after which the cycle inverts the spindle rotation to execute the chip breaking.
Retract	Length of the path traveled by the tap after spindle inversion. Select the option by pressing the SELECT button.
automatically	The cycle automatically sets the length to be traveled.
manual V2	The cycle offers the possibility to set the length of the path manually at parameter V2. Enter the value (e.g. 2 mm).

25.5 Parameters concerning the chip removal

Parameter	Description
D	Length of pass after which the cycle inverts the spindle rotation to bring the tool back outside of the workpiece into the starting position, which is set by the parameter SC.

25.6 Practical exercise

25.6.1 Execution of a rigid axial tapping

Open the program 'PRG_25_01' in the folder 'CHAP_22_27'
This program uses the tapping cycle to create the part shown in the following figure. In the cycle, the option "without compensating chuck" is enabled, i.e. a rigid tapping is executed.
Start the graphic simulation and observe the tool movements. Proceed to change the cycle parameters on the basis of the description given in paragraph 25.3 and observe the variations in the execution of the tooling operation.

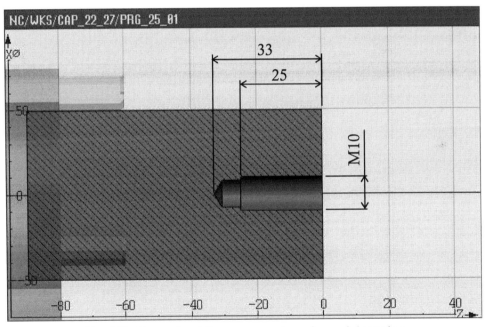

Fig. 166. Example of the programming of an axial tapping

```
; blank part dimensions:
; XA = 50 bar diameter
; ZA = 0 machining allowance on front face
; ZI = -90 length of finished part
; ZB = -60 extension from jaws
N10 WORKPIECE(,,,"CYLINDER",0,0,-90,-60,50)

N20 G18 G54 G90
N30 G0 X400 Z500
```

```
N40 M8
N50 SETMS(1)

N60 T6 D1 ; FIXED AXIAL DRILL D8.5
N70 G95 S1200 M3 F0.1
N80 G0 Z20
N90 G0 X0

N100 CYCLE82(5,0,2,-33,,2,0,1,22)

N110 G0Z200

N120 T7 D1 ; FIXED AXIAL TAPPING TOOL M10
N130 G95 S1200 M3
N140 G0 Z20
N150 G0 X0

N160 CYCLE84(5,0,2,-
25,,0.7,5,,1.5,0,1000,1200,0,1,0,0,5,1.4,,"ISO_METRIC","M10",
,1001,2001002)

N170 G0 Z200
N180 G0 X200
N190 M30
```

25.6.2 Execution of a compensated radial tapping

Open the program 'PRG_25_02' in the folder 'CHAP_22_27' and proceed as in the previous paragraph.

This program uses the tapping cycle to create the part shown in the following figure. In the cycle, the option "with compensating chuck" is enabled, i.e. the tapping is executed with the tool mounted on a holder permitting to the tap to move along the rotating axis.

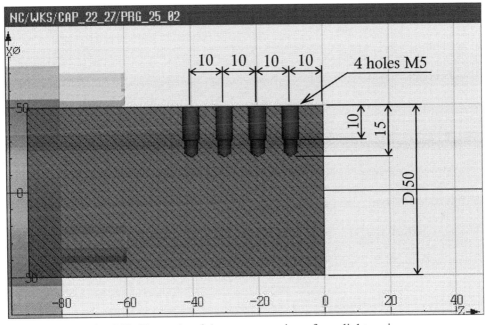

Fig. 167. Example of the programming of a radial tapping

```
; blank part dimensions:
; XA = 50 bar diameter
; ZA = 0 machining allowance on front face
; ZI = -90 length of finished part
; ZB = -60 extension from jaws
N10 WORKPIECE(,,,"CYLINDER",0,0,-90,-60,50)

N20 G18 G54 G90
N30 G0 X400
N40 G0 Z500
N50 M8
N60 SETMS(1)
N70 SPOS=0 ;ANGULAR ORIENTATION OF THE MASTER SPINDLE AT ZERO
DEGREES
```

```
N80 T17 D1 ; DRIVEN RADIAL DRILL D4
N90 SETMS(3) ;SETTING OF THE ROTATING TOOLS AS MASTER
N100 G95 S1800 M3 F0.1 ;FEEDRATE EXPRESSED IN MM/REV
REFERRING TO THE ROTATION OF THE DRIVEN TOOLS
N110 G0 Z-10
N120 G0 X55

N130 G19 ; DEFINITION OF THE WORKING PLANE Y-Z
N140 MCALL CYCLE82(55,50,2,20,,2,0,3,22)
N150 G0 Z-10
N160 G0 Z-20
N170 G0 Z-30
N180 G0 Z-40
N190 MCALL
N200 G18 ; RETURN TO THE TURNING PLANE X-Z
N210 M5 ;DRIVEN TOOLS STOP

N220 G0 X200
N230 G0 Z200

N240 T18 D1 ; DRIVEN RADIAL TAP M5
N250 G95 S1100 M3
N260 G0 Z-10
N270 G0 X55

N280 G19 ; DEFINITION OF THE WORKING PLANE Y-Z
N290 MCALL
CYCLE840(55,50,2,30,,0.7,0,5,11,,0.8,0,1,0,,"ISO_METRIC","M5"
,,1003,2)
N300 G0 Z-10
N310 G0 Z-20
N320 G0 Z-30
N330 G0 Z-40
N340 MCALL
N350 G18 ; RETURN TO THE TURNING PLANE X-Z
N360 M5 ;DRIVEN TOOLS STOP

N370 SETMS(1) ;SETTING OF MAIN SPINDLE AS MASTER
N380 G0 X200
N390 G0 Z200
N400 M30
```

26. CYCLE940: Cycle for Thread Undercuts (1h)
(Theory: 0.5h, Practice: 0.5h)

26.1 Description
This cycle automatically creates the undercut which is often present at the end of a thread. The available parameters in the dialog box allow to determine the shape and the dimension of the undercut according to international rules.

In the cycle, the position of the undercut, its width, the dimension of the rounds to be executed on the profile, the inclination angle with which to enter the material and other values, like the pass depth, the feedrate and the roughing method are defined.

Make sure you execute the tooling operation with a tool which is suitable for the shape to be formed.

26.2 Insertion procedure
Starting from the main menu shown in paragraph 17.3, proceed by pressing the buttons shown in the following table:

 Turning	Horizontal softkey Turning.
 Undercut	Vertical softkey Undercut.
 Undercut thread	Vertical softkey Undercut thread.

Fig. 168. CYCLE940: procedure for the insertion of the cycle

26.3 Parameters

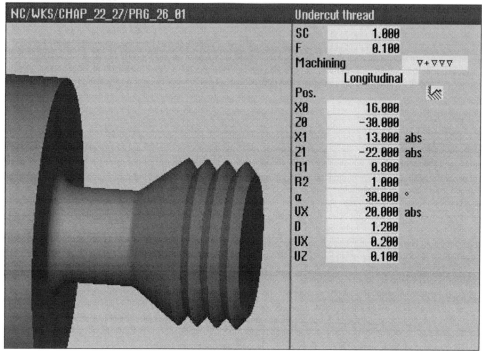

Fig. 169. CYCLE940: window for the insertion of the parameters

SC	Approaching distance. Has diametral value when referring to the X-axes.
F	Feedrate.
Machining	Type of tooling operation to carry out.
▽	Execution of the roughing without elimination of the machining allowance.
▽▽▽	Execution of the finishing pass only.
▽+▽▽▽	Complete execution of the undercut.

Machining	Choice of the trajectory with which to execute the roughing of the profile.
Longitudinal	With longitudinal passes parallel to the Z-axis.
Contour parall.	With passes parallel to the profile of the undercut.
Position	Position of the undercut with respect to the Cartesian axes.
	External oriented in the **negative** Z-axis direction.
	Internal oriented in the **negative** Z-axis direction.
	Internal oriented in the **positive** Z-axis direction.
	External oriented in the **positive** Z-axis direction.
X0	Starting diameter of the undercut.
Z0	Arrival position of the undercut in Z.
X1	End diameter of the undercut.
	Select the option by pressing the SELECT button.
abs	The value of X1 has absolute meaning referring to the part zero point.
inc	The value of X1 has incremental meaning with respect to the starting point expressed by the parameter 'X0'.

CNC – 50 Hour Programming Course

Z1	Starting point of the undercut. Select the option by pressing the SELECT button.
abs	The value of Z1 has absolute meaning referring to the part zero point.
inc	The value of Z1 has incremental meaning with respect to the arrival point expressed by the parameter 'Z0'.
R1	Value of the round radius between the entry cut and the cylindrical part.
R2	Value of the round radius between the cylindrical part and the straight shoulder.
α	Value of the inclination angle of the undercut. Attention: the end cutting edge angle 'α1' needs to be higher than the angle 'α' set in this parameter.
UX	Diameter up to which to execute the straight shoulder at the bottom of the undercut.
D	Pass depth for the execution of the roughing.
UX	Machining allowance of X-axis (radial value).
UZ	Machining allowance of Z-axis.

26.4 Practical exercise

26.4.1 Execution of an undercut for external thread M16

Open the program 'PRG_26_01' in the folder 'CHAP_22_27'
This program uses the cycle CYCLE940 to create the undercut of the thread M16 shown in the following figure.

Start the graphic simulation and observe the tool movements. Proceed to change the cycle parameters on the basis of the description given in paragraph 26.3 and observe the variations in the execution of the tooling operation.

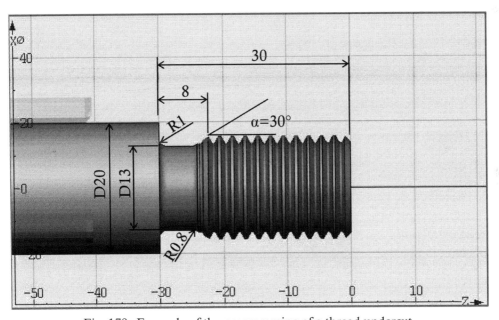

Fig. 170. Example of the programming of a thread undercut

```
; blank part dimensions:
; XA = 20 bar diameter
; ZA = 0.4 machining allowance on front face
; ZI = -86 length of the finished part
; ZB = -41 extension from jaws
WORKPIECE(,,,"CYLINDER",0,0.4,-86,-41,20)

G18 G54 G90
G0 X400 Z500
M8
```

```
SETMS(1)

LIMS=3000
T1 D1 ; EXTERNAL TURNING
G96 S100 M4
G0 X22 Z0
G1 X-1.6 F0.18
G0 X12.8 Z0.5
G1 Z0
G1 X15.8 Z-1.5
G1 Z-30
G1 X22
G0 X200
G0 Z200

T2 D1 ; TURNING TOOL WITH END CUTTING EDGE ANGLE HIGHER THAN 30°
G95 S1200 M4
G0 Z-20
G0 X22

CYCLE940(16,-30,"T",1,1,0.1,13,13,-22,0.8,1,30,20,1.2,0.1,0.2,0.1,,,,2,1000)

G0 X200
G0 Z200

T4 D1 ; TOOL FOR EXTERNAL THREADINGS
G95 S600 M3
G0 Z4
G0 X20

CYCLE99(0,16,-28,,4,0,1.2268,0.05,30,0,5,0,2,1310203,4,2,0.3,0.5,0,0,1,0,0.708293,1,,"ISO_METRIC","M16",1102,0)

G0 X200 ; DISENGAGEMENT
G0 Z200
M30
```

27. Use of the C-Axis (4h)
(Theory: 1h, Practice: 3h)

27.1 Introduction
We now come back to the text already examined in paragraph 4.3, adding new information and programming examples.

27.2 C-axis
The numeric control, thanks to its calculation functions, offers the chance to use the rotating axis of the spindle as an interpolation axis, i.e. it is capable of coordinating its movements on the basis of the movements of the other axes. The rotating axis of the spindle is always coaxial to the Z-axis and is therefore called the C-axis. With the C-axis, it is possible to perform milling and drilling operations, and its use is always associated with the presence of driven tools in the machine.

The C-axis is used in four different ways:
- Angular positioning of the spindle.
- Cylindrical interpolation of lines only on the circumference (simple interpolation Z-C).
- Cylindrical interpolation of profiles containing circular parts (transformed interpolation Z-C).
- Frontal interpolation with virtual Y-axis (transformed interpolation X-C).

Simple interpolation means that the arrival values on the spindle axis are programmed with angular values related to the original characteristics of this axis (rotating axis).
Transformed interpolation means that the spindle axis, which is originally rotating, is transformed into a linear axis; the positionings are programmed by means of linear values (the rotating axis is transformed into a linear axis).

27.2.1 M70: Activation of the C-axis

The C-axis can be used to orientate the spindle in order to perform radial holes coaxial to the X-axis or to drill frontal holes coaxial to the Z-axis.

This usage method overlaps with the SPOS function (see paragraph 8.9.), which remains simpler and faster to program. The advantage is that the angular orientation is already referred to any other tooling operations executed with the C-axis.

Attention: the positive movement of the spindle is opposite to the programming direction, as explained in paragraph 4.4.

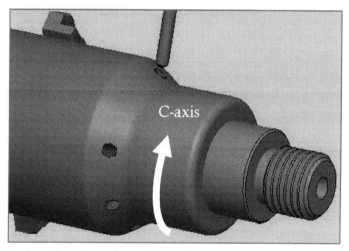

Fig. 171. Angular orientation of the spindle for the creation of radial holes

In lathes designed for processing very large workpieces, the stationing of the spindle in a certain angular position is guaranteed by the presence of a mechanical brake acting directly on a disc combined with the spindle itself. It may be activated by an 'M' function created by the manufacturer.

In smaller machines the absence of the brake shows that the angular orientation and therefore the blocking of the spindle is obtained by keeping the motor of the spindle electrically active. The motor torque itself is the power used to oppose each movement following the tooling operations performed on the workpiece.

The M70 function causes the spindle to function as an axis.
M70 has to be programmed after the wokpiece carrying spindle is defined as master spindle. Its angular positioning is programmed together with the spindle name (in this case SP1).
Go to paragraph 8.2 to review the information on the definition of the reference spindle and the spindle names.
Below is a programming example for the execution of two radial holes at 180 degrees.

```
N10 WORKPIECE(,,,"CYLINDER",0,0,-200,-150,80)

N20 G0 G18 G54 G90
N30 G0 X400 Z500
N40 M8

N50 SETMS(1) ; SETTING OF SPINDLE '1' AS MASTER SPINDLE
N60 M70 ; ACTIVATION OF THE 'C'-AXIS
N70 G0 SP1=0 ; ORIENTATION AT 0° OF THE SPINDLE NAMED SP1

N80 T8 D1; RIGHT HAND RADIAL DRILL D.6
N90 SETMS(3) ; SETTING OF SPINDLE 3 AS MASTER SPINDLE
N100 G95 S1200 M3 ; ROTATION REFERRING TO SPINDLE '3'

N110 G0 Z-40
N120 G0 X84
N130 G1 X50 F0.1
N140 G4 S2
N150 G0 X84

N160 G0 SP1=180 ; ORIENT. AT 180° OF THE SPINDLE NAMED SP1

N170 G1 X50 F0.1
N180 G4 S2
N190 G0 X84
N200 M5 ; DRIVEN TOOLS STOP

N210 SETMS(1)
N220 M5 ; DEACTIVATION OF THE C-AXIS

N230 G0 X200
N240 G0 Z200

N250 M30
```

27.2.2 Milling of lines on the circumference

The second way of using the C-axis allows for the execution of cylindrical interpolations of linear sections on the circumference with simple interpolation Z-C.
The relevant axes are:
- C for rotation (degrees),
- Z for movements on lengths (mm),
- X for the depth of milling on the circumference (mm diametral).

When **milling is performed only on the C-axis**, the feedrate is expressed in degrees per revolution (speed according to the tool rotation) or in degrees per minute (speed independent of number of revolutions). Starting with the feedrate of the tool expressed in millimeters, the corresponding value expressed in degrees is now found.

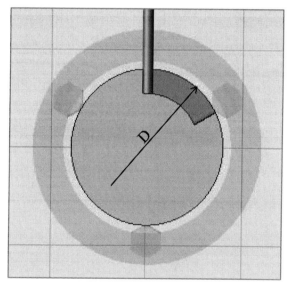

Fig. 172. Calculation of the feedrate expressed in degrees

The feedrate value expressed in degrees must be calculated on the maximum diameter to be worked on, starting with the following proportion:

$$\pi D / 360 = f_{mm} / f_{degrees}$$

The proportion says that the relation between the circumference on the reference diameter and 360 degrees is the same as the relation between the feedrate expressed in millimeters and the feedrate expressed in degrees. From the proportion, the formula for the calculation of the feedrate expressed in degrees is obtained:

$$f_{degrees} = (f_{mm} * 360) / (\pi D)$$

If 'f_{mm}' is expressed in mm/rev., '$f_{degrees}$' will be degrees/rev.
If 'f_{mm}' is expressed in mm/minute, '$f_{degrees}$' will be degrees/minute.

Below is a programming example for the execution of a milling operation at 90 degrees parallel to the direction of the C-axis.
The feedrate in degrees per revolution was calculated starting with a feedrate of 0.1 millimeters per revolution on diameter 80.

```
...
N50  SETMS(1)
N60  M70
N70  G0 SP1=0
N80  T10 D1; RADIAL MILL D.3
N90  SETMS(3)
N100 G95 S3500 M3
N110 G0 Z-20
N120 G0 X82 ; APPROACH TO WORKPIECE
N130 G1 X72 F0.1 ; ENTERING MATERIAL AT DIAM.72
N140 G1 SP1=90 F0.14 ; ROTATION OF 90 DEGREES WITH FEEDRATE
OF 0.14 DEGREES PER REVOLUTION
N150 G1 X82 F1 ; EXIT FROM THE WORKPIECE WITH HIGH FEEDRATE
N160 G0 X200
N170 G0 Z200
...
```

When the **tooling operation is carried out on the C- and the Z-axis simultaneously**, the feedrate is expressed in millimeters and is applied to the Z-axis only (the Z-axis brings the movements of the C-axis into interpolation). **It is the programmer's task to recalculate the feedrate considering the additional shift of the C-axis given by the spindle rotation, according to the formulas shown below.**

The following figure shows a tooling operation obtained by a simple cylindrical interpolation C-Z.

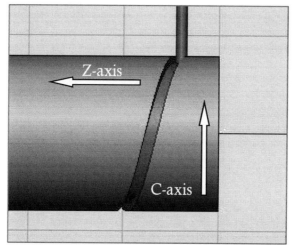

Fig. 173. Calculation of the feedrate in simple cylindrical interpolation C-Z

Calculate the length to be traveled in C according to the following formula:
(D is the reference diameter and α the degrees to be traveled)

$$L_C = \pi D * (\alpha/360)$$

Add the length traveled on Z to the one traveled on C in order to find the total length to be traveled:

$$L_T = L_Z + L_C$$

Take the real feedrate value (f_R) with which you want to make the tool work (e.g. 0.1 mm/rev).
Enter it into the following formula in order to obtain the feedrate to be programmed:

$$f_P = f_R * L_Z / L_T$$

Below is a programming example for the execution of a milling operation at 180 degrees with simultaneous movement of 10 mm on the Z-axis.

The programmed feedrate was calculated as follows:
- D=80 mm: the reference diameter is the external diameter to be milled,
- α=180°: milling length,
- L_Z=10 mm: length of the path to be executed in Z,
- f_R=0.1 mm/rev: real feedrate with which the tool has to work.

$L_C = \pi D * (\alpha/360)$ 3.14*80*180 / 360   = 125,6 mm
$L_T = L_Z + L_C$ 10+125.6   = 135,6 mm
$f_P = f_R * L_Z / L_T$ 0.1*10 / 135.6   = 0.007 mm

```
...
N50 SETMS(1)
N60 M70
N70 G0 SP1=0

N80 T10 D1; RADIAL MILL D.3
N90 SETMS(3)
N100 G95 S1500 M3
N110 G0 Z-20
N120 G0 X82
N130 G1 X72 F0.1
N140 G1 Z-30 SP1=180 F0.007
N150 G1 X82 F1
N160 G0 X200
N170 G0 Z200
...
```

27.2.3 TRACYL: cylindrical interpol. on the circumference

If the profile to be described on the circumference contains arcs or circle arcs, the TRACYL function, which transforms the rotating C-axis into a linear Y-axis to be programmed in millimeters, must be used.

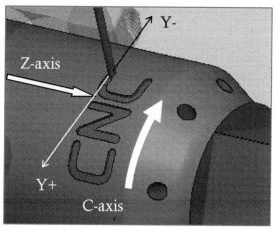

Fig. 174. Interpolation example with TRACYL

The positions in millimeters refer to the circumference on the reference diameter (also called the **interpolation diameter**). The reference diameter is the one on which the programmed profile is executed and it is defined together with the TRACYL function (e.g. TRACYL80).

The following figure shows the positive and negative programming direction of the Y-axis, as the tool moves on the workpiece (the real movement of the spindle is opposite).

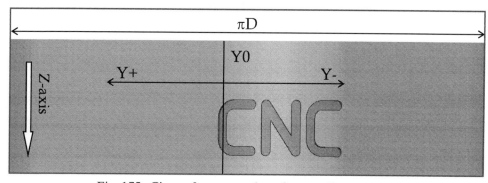

Fig. 175. Circumference on the reference diameter 'D'

Here is the programming example for the execution of the inscription shown in figure 175.

```
N1140 SETMS(1)
N1150 T10 D1; RIGHT HAND RADIAL MILL D3
N1160 S3=2200 M3=3 G94
N1170 G0 X100
N1180 G0 Z-110
N1190 SPOS=0

N1200 TRACYL(80) ; ACTIV. OF CYLINDRICAL INTERP. ON DIAM.80
N1210 G1 X84 Y-10 Z-110 F2000 ; POSITIONING ON THE CARTESIAN
PLANE Y-Z

;LETTER C
N1220 G1 X78 F200
N1230 G1 Y-4
N1240 G2 Y0 Z-114 CR=4
N1250 G1 Z-121
N1260 G2 Y-4 Z-125 CR=4
N1270 G1 Y-10
N1280 G1 X84 F500

N1290 G1 Y-15 Z-110 F2000
;LETTER N
N1300 G1 X78 F200
N1310 G1 Z-125
N1320 G1 Y-25 Z-110
N1330 G1 Z-125
N1340 G1 X84 F500

N1350 G1 Y-40 Z-110 F2000
;LETTER C
N1360 G1 X78 F200
N1370 G1 Y-34
N1380 G2 Y-30 Z-114 CR=4
N1390 G1 Z-121
N1400 G2 Y-34 Z-125 CR=4
N1410 G1 Y-40
N1420 G1 X84 F500

N1430 TRAFOOF ; DISABLING OF TRACYL
N1440 G0 X400
N1450 G0 Z500
```

27.2.4 TRANSMIT: frontal interpolation

With the TRANSMIT function, the interpolation between the C-axis and the X-axis is activated. **It must be used with driven tools coaxial to the Z-axis only.** TRANSMIT permits the numeric control to transform the C-axis into a virtual Y-axis, i.e. an axis transversal to the workpiece, which allows for the execution of out-of-axis millings and any profile described on the frontal plane of the workpiece.

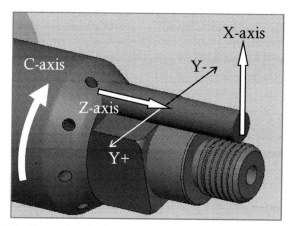

Fig. 176. Example of a frontal interpolation with TRANSMIT

The TRANSMIT function activates a virtual system of Cartesian frontal coordinates like the one shown in figure 177.

The relevant axes are:
- X as real Cartesian axis on which the values are to be expressed in mm,
- Y as virtual Cartesian axis (obtained by the transformation from C) on which the values are to be expressed in mm,
- Z for movements on lengths.

The system of Cartesian axes is to be positioned:
- on the workpiece face,
- with zero point of the axes on the rotating axis of the workpiece as shown in figure 177,
- with angular orientation in the most convenient direction to measure the profile values.

Attention: the values programmed on the X-axes have diametral meaning when the DIAMON function is enabled. It is recommended though to give them radial meaning by programming DIAMOF before the execution of the program.

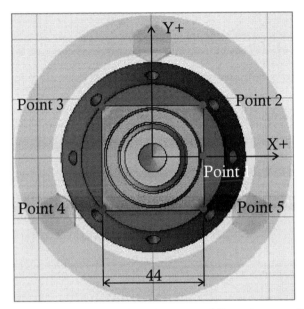

Fig. 177. Position of the virtual system of Cartesian axes Y-X

Below is a programming example for the execution of the square with side of 44 mm shown in figure 177.

```
...
N900 SETMS(1)

N920 T9 D1; RIGHT HAND AXIAL MILL D16
N930 S3=1800 M3=3 G94 F180
N940 G0 X80
N950 G0 Z10

N960 G17 ; DEFINITION OF THE PLANE X-Y
N970 SPOS=0 ; ANGULAR ORIENTATION AT ZERO DEGREES
N980 DIAMOF ; ATTRIBUTION OF RADIAL MEANING TO THE X VALUES
N990 TRANSMIT ; ACTIVATION OF THE FRONTAL INTERPOLATION X-Y

N1000 G0 Z-80 ; POSITIONING IN Z FOR START OF TOOLING OPERATION
```

N1010 G1 X22 Y0 G42 ; POSIT. AT POINT '1' AND ACTIVATION OF THE TOOL RADIUS COMPENSATION WITH TOOL TO THE RIGHT OF THE PROFILE
N1020 G1 Y22 ; ARRIVAL AT POINT '2'
N1030 G1 X-22 ; ARRIVAL AT POINT '3'
N1040 G1 Y-22 ; ARRIVAL AT POINT '4'
N1050 G1 X22 ; ARRIVAL AT POINT '5'
N1060 G1 Y2 ; CLOSING OF PROFILE GOING BEYOND POINT '1'
N1070 G1 X40 G40 ; DISENGAGEMENT AND DEACTIVATION OF THE TOOL RADIUS COMPENSATION

N1080 TRAFOOF ; DISABLING OF TRANSMIT
N1090 G18 ; RETURN TO THE TURNING PLANE X-Z
N1100 DIAMON ; REACTIVATION OF THE DIAMETRAL MEANING ATTRIBUTED TO THE VALUES PROGRAMMED ON THE X-AXIS

N1110 G0 Z200
N1120 G0 X200
...

27.3 Practical exercise

27.3.1 Angular orientation and milling on the circumference

Open the program 'PRG_27_01' in the folder 'CHAP_22_27' This program executes the workpiece shown in figure 178. It uses the C-axis for the angular orientation of the spindle (for the execution of the radial hole) and to execute a simple cylindrical interpolation Z-C (for the milling on the circumference).

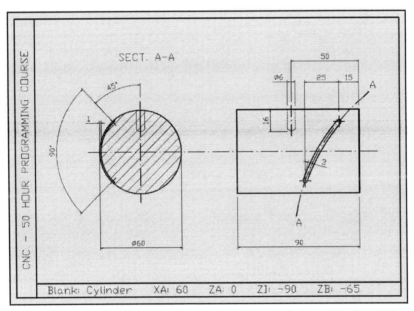

Fig. 178. Example of the programming of a simple interpolation Z-C

The programmed feedrate was calculated as follows:
- (D=60 mm): the reference diameter is the external diameter to be milled,
- (α=90°): milling length,
- (L_Z=25mm): length of the path to be executed in Z,
- (f_R=0.1 mm/rev): real feedrate with which the tool has to work.

$$L_C = \pi D * (\alpha/360) \quad 3.14*60*90 / 360 \quad = 47,1mm$$
$$L_T = L_Z + L_C \quad 25+47.1 \quad = 72,1mm$$
$$f_P = f_R * L_Z / L_T \quad 0.1*25 / 72.1 \quad = 0.03 degrees/rev$$

The following program executes the hole and the milling shown in figure 178.

```
; blank part dimensions:
; XA = 60 bar diameter
; ZA = 0 machining allowance on front face
; ZI = -90 length of finished part
; ZB = -65 extension from jaws

N10 WORKPIECE(,,,"CYLINDER",0,0,-90,-65,60)

N20 G0 G18 G54 G90
N30 G0 X400 Z500
N40 M8

N50 SETMS(1)
N60 M70
N70 G0 SP1=0 ; ANGULAR POSITIONING OF THE SPINDLE AT 0° FOR THE EXECUTION OF THE HOLE

N80 T8 D1; RADIAL DRILL D.6
N90 SETMS(3)
N100 G95 S1100 M3
N110 G0 Z-50
N120 G0 X62
N130 G1 X28 F0.12
N140 G4 S2 ; DWELL TIME OF TWO REVOLUTIONS AT BOTTOM OF HOLE
N150 G0 X200

N160 T10 D1; RADIAL MILL D.3
N170 SETMS(3)
N180 G95 S1500 M3
N190 G0 SP1=45 ; ANGULAR POSITIONING OF THE SPINDLE AT 45° AS STARTING ANGLE FOR THE TOOLING OPERATION
N200 G0 Z-15 ; POSITIONING OF THE TOOL IN Z
N210 G0 X62 ; RAPID APPROACH TO WORKPIECE IN X
N220 G1 X58 F0.1 ; WORKING MOVEMENT TO THE BOTTOM OF THE MILLING
N230 G1 Z=IC(-25) SP1=IC(90) F0.035 ; EXECUTION OF THE MILLING IN INCREMENTAL COORDINATES WITH RESPECT TO THE STARTING POINT
N240 G1 X62 F1
N250 G0 X200
N260 G0 Z200

N270 M30
```

27.3.2 Cylindrical interpolation with TRACYL

Open the program 'PRG_27_02' in the folder 'CHAP_22_27'
This program executes the workpiece shown in figure 179.

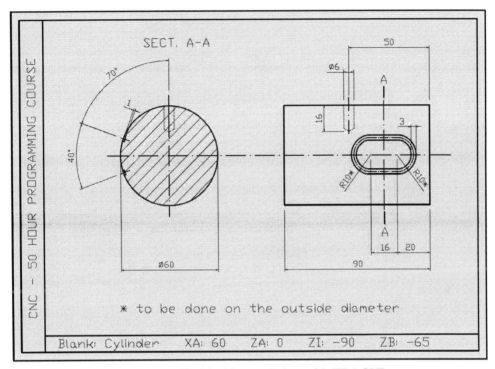

Fig. 179. Cylindrical interpolation with TRACYL

The milling on the circumference refers to the position of the radial hole. The presence of radii in the profile shows that it is necessary to use the TRACYL function. The most common difficulty in the execution of these workpieces is that the drawing normally shows the values in degrees, while the programs need values in millimeters.

Proceed as follows: the distance between the hole and the first milling line is 70 degrees. The note on the drawing says that the radius dimension refers to the external diameter of the workpiece (60 mm). After activation of the TRACYL function, it is however necessary to express those values in millimeters, referring them to the circumference on the reference diameter.

The following figure shows how the TRACYL function interprets the new reference system.

The values expressed in millimeters were calculated by means of the formula already used in paragraph 27.2.2, where D is the reference diameter and α shows the position in degrees.

$$L_C = \pi D * (\alpha/360)$$

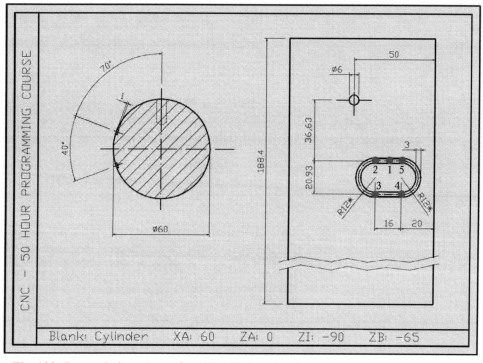

Fig. 180. Interpolation plane after TRACYL activation (reference diameter 60 mm)

The programmed feedrate on X and Z is applied to the movements on Y as well and is calculated by the NC on the reference diameter.

On the next page you will find the program for the execution of the workpiece shown in figure 179.

```
; blank part dimensions:
; XA = 60 bar diameter
; ZA = 0 machining allowance on front face
; ZI = -90 length of finished part
; ZB = -65 extension from jaws
N10 WORKPIECE(,,,"CYLINDER",192,0,-90,-65,60)

N20 G0 G18 G54 G90
N30 G0 X400 Z500
N40 M8
N50 SETMS(1)
N60 M70
N70 G0 SP1=0

N80 T8 D1; RADIAL DRILL D.6
N90 SETMS(3)
N100 G95 S1100 M3
N110 G0 Z-50
N120 G0 X62
N130 G1 X28 F0.12
N140 G4 S2
N150 G0 X200

N160 T10 D1; RADIAL MILL D.3
N170 SETMS(3)
N180 G95 S1500 M3
N190 G0 Z-28
N200 G0 X62

N210 G19
N220 TRACYL(60)

N230 G0 Y36.63
N240 G1 X58 F0.1
N250 G1 Z-36 F0.1
N260 G3 Y=IC(20.93) Z-36 CR=12
N270 G1 Z-20
N280 G3 Y=IC(-20.93) Z-20 CR=12
N290 G1 Z-30
N300 G1 X62

N310 TRAFOOF
N320 G18

N330 G0 X200
N340 G0 Z200
N350 M30
```

27.3.3 Frontal interpolation with TRANSMIT

Open the program 'PRG_27_03' in the folder 'CHAP_22_27'
This program executes the workpiece shown in figure 181 using the TRANSMIT function.

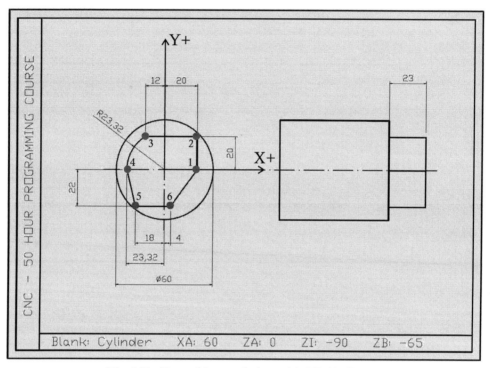

Fig. 181. Frontal interpolation with TRANSMIT

The profile as the one shown here can be of any type and may contain lines, inclined parts and radii.

```
; blank part dimensions:
; XA = 60 bar diameter
; ZA = 0 machining allowance on front face
; ZI = -90 length of finished part
; ZB = -65 extension from jaws

N10 WORKPIECE(,,,"CYLINDER",0,0,-90,-65,60)

N20 G0 G18 G54 G90
N30 G0 X400 Z500
N40 M8
```

```
N50 SETMS(1)
N60 M70

N70 T9 D1; AXIAL MILL D.16
N80 S3=1800 M3=3 G94 F180
N90 G0 Z-23 ; POSITIONING OF THE TOOL IN Z
N100 G0 X84 ; RAPID APPROACH TO WORKPIECE IN X

N110 G17 ; DEFINITION OF THE WORKING PLANE X-Y
N120 SPOS=0 ; ANGULAR ORIENTATION AT ZERO DEGREES
N130 DIAMOF ; DISABLING OF THE DIAMETERAL VALUE OF THE X-AXES

N140 TRANSMIT ; ACTIVATION OF THE FRONTAL INTERPOLATION
N150 G1 X20 Y0 F0.12 G42 ; POSITIONING AT POINT 1
N160 G1 Y20 ; ARRIVAL AT POINT '2'
N170 G1 X-12 ; ARRIVAL AT POINT '3'
N180 G3 Y0 X-23.32 CR=23.32 ; ARRIVAL AT POINT '4'
N190 G1 X-18 Y-22 ; ARRIVAL AT POINT '5'
N200 G1 X4 ; ARRIVAL AT POINT '6'
N210 G1 X20 Y0 ; ARRIVAL AT POINT '1'
N220 G1 X20 Y2 ; CLOSING OF PROFILE
N230 G1 X40 G40 F1 ; DISENGAGEMENT
N240 TRAFOOF ; DISABLING OF TRANSMIT
N250 DIAMON ; ENABLING OF THE DIAMETERAL VALUE OF THE X-AXES

N260 G0 X200
N270 G0 Z200
N280 M30
```

28. Third Test (2h)
(Practice: 2h)

28.1 Introduction to the test
The test consists in the execution of the program creating the part shown in figure 183. Take the following steps:
- Load the tool files which you find in folder 01_EXERCISES named EMPTY_TOOL_LIST. This file deletes all the existing tools by overwriting them only with the roughing tool defined therein. In order to load this file, follow the procedure laid down in paragraph 3.3.
- Create the necessary tools for the execution of this program following the procedure specified in paragraph 7.5.1.
- Create an empty main program in the folder 01_EXERCISES and name it TEST_28_01.
- Structure the program as the ones we've seen so far:

28.2 Tooling operations and cutting parameters

Tooling sequence	Tool	Operation	Cutting speed (m/min)	Feedrate (mm/rev)
Tooling operation no. 1: use the roughing cycle				
1st	T1 D1	External roughing	100	0.18
Tooling operation no. 2: program the finished profile by using ISO codes				
2nd	T2 D1	External finishing	120	0.12

Tooling sequence	Tool	Operation	Cutting speed (m/min)	Feedrate (mm/rev)
Tooling operation no. 3: use the cycle for undercuts, define the insert angle at 35°				
3rd	T2 D1	Undercut	100	0.12
Tooling operation no. 4: use the threading cycle				
4th	T4 D1	Threading M36x4	60	-
Tooling operation no. 5: use the ISO codes, execute center drilling up to Z-4				
5th	T5 D1	Center drilling	80	0.08
Tooling operation no. 6: use the drilling cycle				
6th	T6 D1	Hole D8.5	80	0.1
Tooling operation no. 7: use the rigid tapping cycle				
7th	T7 D1	Tap M10	40	-
Tooling operation no. 8: use the function M70, calculate the feedrate in degrees per revolution				
8th	T10 D1	Milling on circumference	35	0.06
Tooling operation no. 9: use the TRANSMIT function				
9th	T9D1	Frontal milling	80	0.1

Fig. 182. Sequence of tooling operations and cutting parameters to use for the test

28.3 Drawing of the part to be created

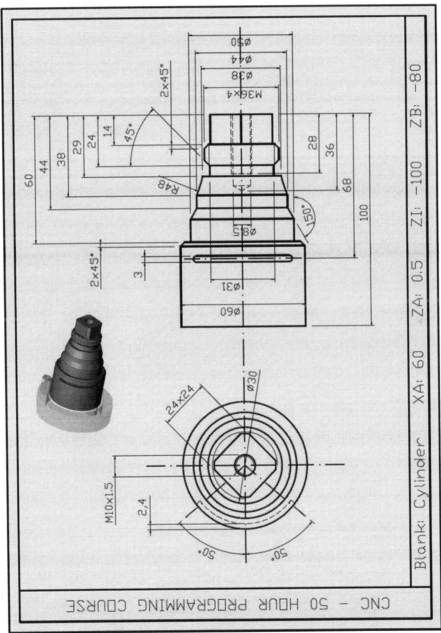

Fig. 183. Drawing of the part to be created

28.4 Support program for the test

For the test, you can make use of the following program which creates the part shown in figure 184 and which has already been examined in chapter 3.

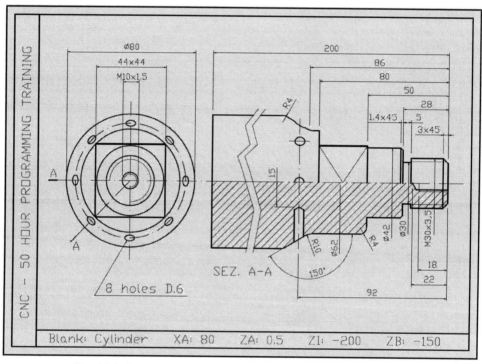

Fig. 184. Part created by the support program for the 3rd test

```
; blank part dimensions:
; XA = 80 bar diameter
; ZA = 0.5 machining allowance on front face
; ZI = -200 length of finished part
; ZB = -150 extension from jaws

N10 WORKPIECE(,,,"CYLINDER",192,0.5,-200,-150,80)

N20 G18 G54 G90
N30 G0 X400 Z500
N40 M8
N50 SETMS(1)

N60 T1 D1 ; LEFT HAND ROUGHING TOOL
N70 G95 S1800 M4
```

```
N80 G0 X82 Z0
N90 G1 X6 F0.2
N100 G0 X80 Z2

N110 CYCLE62(,2,"PROFILE1","END1")

N120
CYCLE952("temp_con",,"",2101311,0.1,0,0,3,0.1,0.1,0.5,0.1,0.1
,0,1,0,0,,,,,,2,2,,,0,1,,0,12,1100110)

N130 G0 X200 Z200

N140 T2 D1 ; LEFT HAND FINISHING TOOL
N150 G95 S2400 M4
N160 G0 X22 Z2
N170 PROFILE1:
N180 G0 X24 Z0
N190 G1 X30 ANG=135 F0.12
N200 G1 Z-28
N210 G1 X42 CHF=1.4
N220 G1 Z-50
N230 G1 X62 RND=4
N240 G1 Z-86 RND=10
N250 G1 X80 ANG=150 RND=4
N260 G1 Z-106
N270 END1:
N280 G0 X200 Z200

N290 T3 D1 ; GROOVING TOOL 3MM
N300 G95 S1200 M4
N310 G0 X34 Z-26
N320 G1 X32 Z-28 F0.3
N330 G1 X25 F0.08
N340 G4 S2
N350 G1 X32 F0.5
N360 G1 Z=IC(2)
N370 G1 X25 F0.08
N380 G4 S2
N390 G1 X32 F0.3
N400 G0 X200 Z200

N410 T4 D1; EXTERNAL THREAD M30
N420 G95 S900 M3
N430 G0 X34 Z2
```

```
N440 CYCLE99(0,30,-
27.8,,6,0,2.1469,0,30,0,8,0,3.5,1310103,4,2,0.3,0.5,0,0,1,0,0
.866,1,,"ISO_METRIC","M30",1102,1001)

N450 G0 X200 Z400

N460 T5 D1; RIGHT HAND CENTER DRIL
N470 G95 S1600 M3
N480 G0 X0 Z2
N490 G1 Z-6 F0.08
N500 G4 S2; DWELL TIME OF 2 REVOLUTIONS ON THE ARRIVAL POINT
N510 G0 Z200

N520 T6 D1; RIGHT HAND AXIAL DRILL D.8.5
N530 G95 S950 M3
N540 G0 X0 Z2
N550 G1 Z-22
N560 G4 F0.1; DWELL TIME OF 0.1 SECONDS AT THE BOTTOM OF HOLE
N570 G0 Z200

N580 T7 D1; TAP M10x1.5
N590 G95 S800 M3
N600 G0 X0 Z4
N610 CYCLE84(4,0,2,-
18,,0.2,3,,1.5,0,800,1200,0,1,0,0,,1.4,,"ISO_METRIC","M10",,1
001,1001002)
N620 G0 Z200
N630 G0 X200

N640 SPOS=0
N650 SETMS(3)

N660 T8 D1; RIGHT HAND RADIAL DRILL D.6
N670 G95 S1200 M3
N680 G0 Z-92
N690 STR_HOLE1:
N700 G0 X74
N710 G1 X30
N720 G4 S2
N730 G0 X74
N740 END_HOLE1:

N750 SPOS[1]=45
N760 REPEAT STR_HOLE1 END_HOLE1
N770 SPOS[1]=90
N780 REPEAT STR_HOLE1 END_HOLE1
N790 SPOS[1]=135
```

```
N800  REPEAT STR_HOLE1 END_HOLE1
N810  SPOS[1]=180
N820  REPEAT STR_HOLE1 END_HOLE1
N830  SPOS[1]=225
N840  REPEAT STR_HOLE1 END_HOLE1
N850  SPOS[1]=270
N860  REPEAT STR_HOLE1 END_HOLE1
N870  SPOS[1]=310
N880  REPEAT STR_HOLE1 END_HOLE1
N890  G0 X200
N900  G0 Z200

N910  SETMS(1)

N920  T9 D1; RIGHT HAND AXIAL MILL D16
N930  S3=1800 M3=3 G94 F180
N940  G0 X80
N950  G0 Z10
N960  G17
N970  SPOS=0
N980  DIAMOF
N990  TRANSMIT
N1000 G0 Z-80
N1010 G1 X22 Y0 G42 ;1
N1020 G1 Y22 ;2
N1030 G1 X-22 ;3
N1040 G1 Y-22 ;4
N1050 G1 X22 ;5
N1060 G1 Y2; 6
N1070 G1 X40 G40
N1080 TRAFOOF
N1090 G18
N1100 DIAMON
N1110 G0 Z200
N1120 G0 X200

N1130 SETMS(1)

N1140 T10 D1; RADIAL MILL D3 FOR RIGHT HAND INSCRIPTION
N1150 S3=2200 M3=3 G94
N1160 G0 X100
N1170 G0 Z-110
N1180 SPOS=0

N1190 TRACYL(80)
N1200 G1 X84 Y-10 Z-110 F2000
```

```
;LETTER C
N1210 G1 X78 F200
N1220 G1 Y-4
N1230 G2 Y0 Z-114 CR=4
N1240 G1 Z-121
N1250 G2 Y-4 Z-125 CR=4
N1260 G1 Y-10

N1270 G1 X84 F500
N1280 G1 Y-15 Z-110 F2000

;LETTER N
N1290 G1 X78 F200
N1300 G1 Z-125
N1310 G1 Y-25 Z-110
N1320 G1 Z-125

N1330 G1 X84 F500
N1340 G1 Y-40 Z-110 F2000

;LETTER C
N1350 G1 X78 F200
N1360 G1 Y-34
N1370 G2 Y-30 Z-114 CR=4
N1380 G1 Z-121
N1390 G2 Y-34 Z-125 CR=4
N1400 G1 Y-40
N1410 G1 X84 F500

N1420 TRAFOOF
N1430 G0 X200
N1440 G0 Z200

N1450 M30
```

28.5 Program correction and reloading of tool list

Compare your program to the one in the folder FINISHED_EXERCISES named TEST_28_01.

Before reading the next chapter, reload the complete tool list contained in the folder 01_EXERCISES.

29. CYCLE60: Engraving Cycle (1h)
(Theory: 0.5h, Practice: 0.5h)

29.1 Description
This practical cycle allows for the execution of engravings on the circumference (on plane G19 with cylindrical interpolation Y-Z) or on the front face of the workpiece (on plane G17 with frontal interpolation X-Y). With the parameters shown in the dialog box it is able to control the orientation of the inscription in a simple and intuitive way.

Fig. 185. 1: engraving on plane G19, 2: engraving on plane G17

29.2 Insertion procedure
Starting from the main menu shown in paragraph 17.3, proceed by pressing the buttons shown in the following table:

Milling	Horizontal softkey Milling.
Engraving	Vertical softkey Engraving.

Fig. 186. CYCLE60: procedure for the insertion of the cycle

29.3 Parameters (cylindrical interpolation)

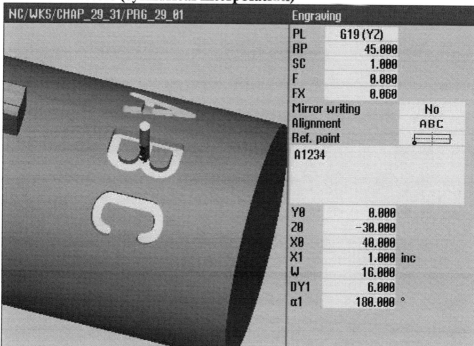

Fig. 187. CYCLE60: window for the insertion of the parameters

Parameter	Description
PL	Setting of the plane on which to execute the engraving. Choose G19 to write on the circumference.
RP	Radial value for the disengagement after the tooling operation.
SC	Incremental approach distance (with radial value) referring to the starting point (parameter X0).
F	Feedrate during engraving.

FX	Feedrate during penetration along the X-axis.
Mirror writing	Executes a mirrored engraving. This parameter is very useful in machines where the positive direction of the C-axis is opposite to the standard direction.
Yes	Activation of the mirrored writing.
No	Deactivation of the mirrored writing.
Alignment	Writing of the text as a line or an arc.
Ref. point	The position of the reference point with respect to the engraving (its position on the circumference is defined by the parameters Y0 and Z0).
Engraved text	This is where the text to be engraved is entered.
Y0	Position in Y of the reference point.
Z0	Position in Z of the reference point.
X0	Radial reference point on which the engraving is carried out. **This point, expressed with diametral value, is to be associated with the TRACYL function before the cycle activation.**
X1	Depth of the engraving with respect to parameter X0.

W	Height of the letter expressed in millimeters.
DY1	Space between the letters expressed in millimeters.
α1	Orientation of the text. **With a value of 90° the text is engraved in parallel direction to the Z-axis.** **With a value of 180° the text is upside down with respect to the front face of the workpiece.**

29.4 Parameters (frontal interpolation)

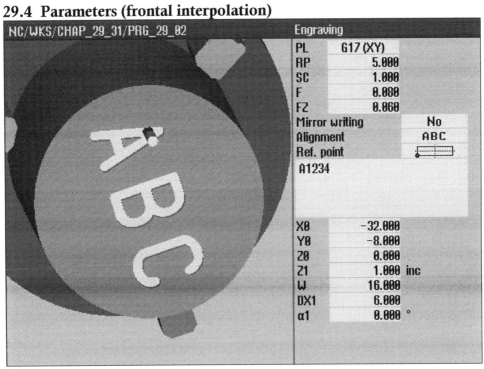

Fig. 188. CYCLE82: window for the insertion of the parameters

Parameter	Description
PL	Setting of the plane on which to execute the engraving. Choose G17 to write on the face of the workpiece.
RP	Value for the disengagement after the tooling operation.
SC	Incremental approach distance referring to the starting point (parameter Z0).
F	Feedrate during engraving.

FX	Feedrate during penetration along the Z-axis.
Mirror writing Yes No	Executes a mirrored engraving. This parameter is very useful when the positive direction of the C-axis is opposite to the standard direction. Activation of the mirrored writing. Deactivation of the mirrored writing.
Alignment	Writing of the text as a line or an arc.
Ref. point	The position of the reference point with respect to the engraving (its position on the part face is defined by the parameters X0 and Y0).
Engraved text	This is where the text to be engraved is entered.
X0	Position in X of the reference point.
Y0	Position in Y of the reference point.
Z0	Position in Z of the plane on which to execute the engraving.
Z1	Depth of the engraving with respect to parameter Z0.

W	Height of the letter expressed in millimeters.
DX1	Space between the letters expressed in millimeters.
α1	Orientation of the text.

29.5 Practical exercise

29.5.1 Example of an engraving on the circumference
Open the program 'PRG_29_01' in the folder 'CHAP_29_31'
This program uses the engraving cycle to write a generic text on the circumference of the workpiece.
Start the graphic simulation and proceed with changing the cycle parameters on the basis of the description given in paragraph 29.3.

Attention: before programming the cycle it is necessary to activate the TRACYL function followed by the diameter on which you wish to engrave the text.

Fig. 189. Use of the engraving cycle to write on the circumference

```
; blank part dimensions:
; XA = 80 bar diameter
; ZA = 0 machining allowance on front face
; ZI = -150 length of finished part
; ZB = -100 extension from jaws
N10 WORKPIECE(,,,"CYLINDER",0,0,-150,-100,80)

N20 G18 G54 G90
```

```
N30 G0 X400 Z500
N40 M8

N50 SETMS(3)
N60 T10 D1 ; RADIAL MILL D3
N70 G95 S4200 M3 F0.06
N80 G0 Z-30
N90 TRACYL(80)

N100 CYCLE60("A1234",45,40,1,,1,0,-
30,180,0,0,16,6,0.06,0.08,20000000,1252,0,100,13,1)

N110 TRAFOOF

N120 G0 X200 Z200
N130 M30
```

29.5.2 Example of a front face engraving

Open the program 'PRG_29_02' in the folder 'CHAP_29_31'
This program uses the engraving cycle to write a generic text on the front face of the workpiece.
Start the graphic simulation and proceed with changing the cycle parameters on the basis of the description given in paragraph 29.4.

Attention: before programming the cycle it is necessary to enable the TRANSMIT function to activate the frontal interpolation X-Y.

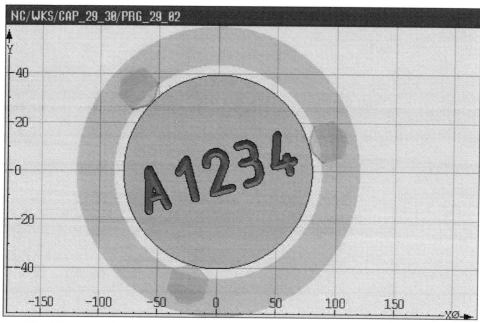

Fig. 190. Use of the engraving cycle to write on the front face of the workpiece

```
; blank part dimensions:
; XA = 80 bar diameter
; ZA = 0 machining allowance on front face
; ZI = -150 length of finished part
; ZB = -100 extension from jaws
N10 WORKPIECE(,,,"CYLINDER",0,0,-150,-100,80)

N20 G18 G54 G90
N30 G0 X400 Z500
N40 M8
```

```
N50 SETMS(3)
N60 T5 D1 ; AXIAL CENTER DRILL
N70 G95 S4200 M3 F0.06
N80 G0 Z10
N90 TRANSMIT

N100 CYCLE60("A1234",5,0,1,,1,-32,-
8,0,0,0,16,6,0.06,0.08,20000000,1252,0,100,11,1)

N110 TRAFOOF

N120 G0 X200 Z200
N130 M30
```

To see the programmed engraving, wait for the generation of the image in high resolution at the end of the graphic simulation.

30. Parametric Programming (2h)
(Theory: 1h, Practice: 1h)

30.1 Use of parametric programming

Parametric programming, also called programming with variables, consists in substituting the numeric value of a word with a variable (G0X100 → R0=100 → G0X=R0), the value of which can be defined in the program or manually in the respective page.

It offers the possibility to automatically update a program based on the numeric value of the variables contained within it. This is useful in the creation of a program for the production of similar parts (belonging to the same family), where the dimensions of the part to be produced are set by the programmer by means of values associated with the variables.

With the variables it is furthermore possible to determine program skips based on their value. Thereby, a logical scheme, called an algorithm, is defined, which modifies the sequence for the execution of the tooling operations contained in the program.

30.2 Calculation variables 'R'

The numeric control offers 100 calculation variables whose value is set directly by the operator or calculated by means of the programmed formula.

The calculation variables are also called 'R' variables as they are identified by the address 'R' in combination with a number between 0 and 99. The 'R' variables are 'real' variables, i.e. they may only have a real number (e.g. 121.074) as a value and not a name or a letter. They are contained on a specific page of the numeric control (see figure 191) and may be accessed following the procedure shown in figure 192.

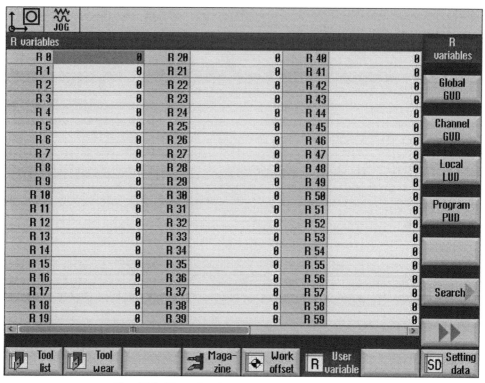

Fig. 191. Page for the calculation variables 'R'

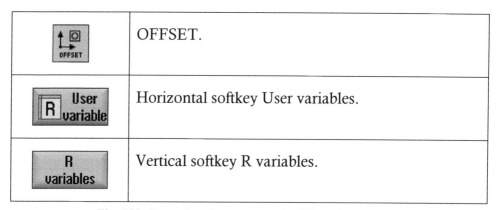

Fig. 192. Procedure for access to the 'R' variables page

30.3 System variables

The system variables express values which are generally linked to the machine data.

They are defined and made available by the numeric control, they can be processed in the program both in reading and in writing and they enable access to zero point offsets, tool corrections, position values of axes and statuses of the numeric control.

The name of the system variables always starts with the symbol $, followed by two letters.

The first letter after $ stands for:

$M	Machine data.
$S	Setting data.
$T	Tool management data.
$P	Programmed values.
$A	Current values.
$V	Service data.

Fig. 193. Meaning of the first letter in the name of system variables

The second letter after $ stands for:

N	Global NCK.
C	Specification for channels.
A	Specification for axis.

Fig. 194. Meaning of the second letter in the name of system variables

The list of system variables can be found in Siemens manuals.

30.4 Symbols for arithmetic calculations
The following symbols allow for the execution of arithmetic operations.

+	Sum.
-	Subtraction.
*	Multiplication.
/	Division.

Fig. 195. Arithmetic functions

Programming syntax:

```
Sum:             R2=R0+R1
                 R0=10+15
Subtraction:     R2=R0-R1
                 R0=22-13
Multiplication:  R14=R8*2
                 R99=16*4
Division:        R50=R14/2
                 R51=R22/R0
```

The expressions are programmed according to common arithmetic rules:

```
R0=(10+2)/2         ; RESULT EQUAL TO 6
R0=10+(2/2)         ; RESULT EQUAL TO 11
R0=(16+2)/(7-5)     ; RESULT EQUAL TO 9
```

30.5 Symbols for trigonometric calculations
The following functions allow for the execution of trigonometric operations.

SIN(…)	Sine of the angle.
COS(…)	Cosine of the angle.
TAN(…)	Tangent of the angle.
ASIN(…)	Arcsine.

ACOS(...)	Arccosine.
ATAN2(1°, 2°)	Arctangent with argument arcsine (1°) and argument arccosine (2°).

Fig. 196. Trigonometric functions

Programming syntax:
```
Sine:         R0=SIN(30)      ; R0=0.5
Cosine:       R0=COS(30)      ; R0=0.866
Tangent:      R0=TAN(45)      ; R0=1
Arcsine:      R0=ASIN(0.5)    ; R0=30
Arccosine:    R0=ACOS(0.866)  ; R0=30
Arctangent2:  R0=ATAN2(1,1)   ; R0=45
```

30.5.1 Calculation scheme of the trigonometric functions

From the Greek 'trigonon' (triangle) and 'metron' (measure), trigonometry offers the formulas to calculate the dimensions which characterize the elements of a triangle, which in this case needs to be rectangle.

Below a summarizing scheme of the trigonometric formulas examined in the previous paragraph.

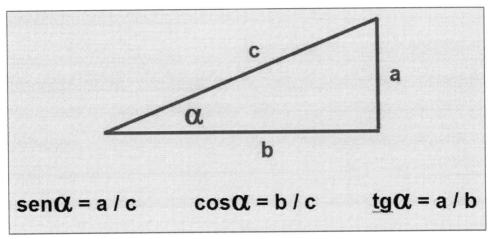

Fig. 197. Calculation scheme of the trigonometric functions

30.6 Result management instructions

The following instructions allow for the management of rounding off a result.

TRUNC	Whole part.
ROUND	Rounding to whole number.

Fig. 198. Result management functions

Programming syntax:
```
Whole part        R0=TRUNC(10.52)    ; R0=10
Rounding:         R0=ROUND(10.52)    ; R0=11
```

The TRUNC function takes the sole whole part of the associated number.

The ROUND function rounds the associated number to the nearest whole number.

30.7 Practical exercise

30.7.1 Calculation test in operating mode MDA

Press the MDA button on the control panel in order to select the operating mode for manual data entry.
Insert the single blocks specified after the figure and press the green CYCLE START button to execute the programmed instruction.
Check the result of the calculation in the page for the user variables.
Go back to the MDA page and press the RESET button to continue with the data entry.

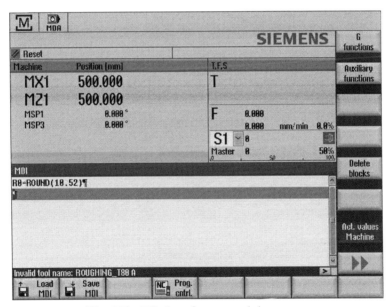

Fig. 199. Window for manual data entry

```
Expression:      R0=(10+2)/2         ; R0=6
Expression:      R0=10+2/2           ; R0=11
Expression:      R0=(16+2)/(7-5)     ; R0=9
Sine:            R0=SIN(30)          ; R0=0.5
Cosine:          R0=COS(30)          ; R0=0.866
Tangent:         R0=TAN(45)          ; R0=1
Arcsine:         R0=ASIN(0.5)        ; R0=30
Arccosine:       R0=ACOS(0.866)      ; R0=30
Arctangent2:     R0=ATAN2(1,1)       ; R0=45
Whole part       R0=TRUNC(10.52)     ; R0=10
Rounding:        R0=ROUND(10.52)     ; R0=11
```

30.7.2 Programming of a workpiece family

A workpiece family is a group of workpieces with similar characteristics as far as their dimensions are concerned and which have differences which repeat themselves and which are unmistakably identifiable.

In this exercise, a program (PRG_30_01 in the folder CHAP_29_31) for the creation of the workpiece family described in the following figure is analyzed. The variations in the dimensions are shown in the table in the drawing.

The program uses variables for describing the data which change from one workpiece to another belonging to the same family.

A very interesting aspect of the exercise is the use of the variables in the working cycles of this program.

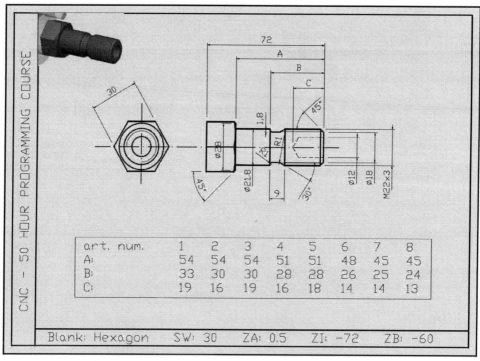

Fig. 200. Drawing of a workpiece family

```
; blank part dimensions:
; blank part: Polygon
; N = 6 number of edges
; SW = 30 hexagon key
; ZA = 0.5 machining allowance on front face
```

```
; ZI = -72 length of finished part
; ZB = -60 extension from jaws

N10 WORKPIECE(,,,"N_CORNER",0,0.5,-72,-60,6,30)
```

At the beginning of the program, a value is attributed to the variable on the basis of the dimensions of the part to be produced.

```
N20 R0=54 ; LENGTH OF TURNING (A)
N30 R1=33 ; THREAD SHOULDER (B)
N40 R2=19 ; HOLE DEPTH (C)

N50 G18 G54 G90
N60 G0 X400 Z500
N70 M8
N80 SETMS(1)

N90 T1 D1 ; TURNING TOOL
N100 G95 S2000 M4 F0.1
N110 G0 X36 Z0
N120 G1 X-1
N130 G0 X28 Z0.2
```

The Z arrival value of the roughing pass is expressed by means of the variable R0, minus the machining allowance to be removed in the finishing pass.

```
N140 G1 Z=-(R0-0.1)
N150 G0 X30 Z0.2
N160 G0 X18
N170 G1 Z0
N180 G1 X21.8 ANG=135
N190 G1 Z=-R0
N200 G1 X28
N210 G1 X36 ANG=135

N220 G0 Z=-(R1-9)
```

Note in figure 201 how the variable R1 is used inside the cycle as well for the execution of the thread undercut.

```
N230 CYCLE940(22,-R1,
"T",1,1,0.1,13,1.8,9,1,1,30,36,1.2,0.1,0.2,0.1,,,,2,1011)
```

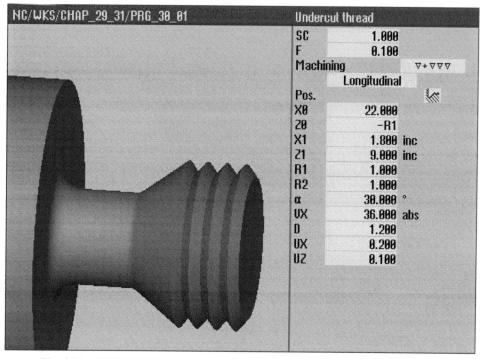

Fig. 201. CYCLE940 with parameter defined by means of the variable R1

```
N240 G0 X300 Z200

N250 T4 D1 ; TOOL FOR EXTERNAL THREADINGS
N260 G95 S1400 M3
N270 G0 X26 Z6
```

Note in figure 202 how the variable R1 is used in the threading cycle in a formula for the execution of an automatic calculation; the arrival point of the thread has been brought forward by 0.5 mm with respect to the shoulder.

```
N280 CYCLE99(0,22,-(R1-0.5),
,6,0,1.5335,0,30,0,8,0,2.5,1310103,4,2,0.3,0.5,0,0,1,0,0.866,
1,,"ISO_METRIC","M22",1102,1000)
```

The following figure shows the dialog box for the insertion of the parameters relative to the creation of the thread.

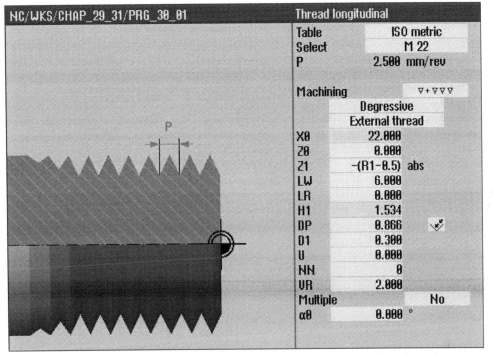

Fig. 202. CYCLE99 with parameter defined by the variable R1 in a formula

```
N290 G0 X300 Z200

N300 T5 D1 ; CENTER DRILL
N310 G95 S2200 M3 F0.06
N320 G0 X0 Z2
N330 G1 Z-4
N340 G0 Z200

N350 T16 D1; AXIAL DRILL D12
N360 G95 S1250 M3
N370 G0 Z2 X0
```

The length of the drill is expressed by means of the variable R2.

```
N380 G1 Z=-R2
N390 G4 S2

N400 G0 Z200
N410 G0 X300
N420 M30
```

For the execution of the other parts belonging to the same family it is sufficient to update the value of the variables at the beginning of the program.

Now create the remaining parts belonging to the same family.

31. Conditional Program Skips (2h)
(Theory: 1h, Practice: 1h)

31.1 The algorithm
Conditional skips are the programmed expressions which permit modification of the linear execution of a program based on the value attributed to one or more variables.
The criterion according to which the execution of a program is modified is called an algorithm.
The algorithm is the logical scheme, defined in the program, which determines the operation sequence within the program.
The algorithm is created by the programmer with the aim to carry out a task (creating 12 holes) or resolving a problem (stopping the program to change the tool after a certain number of pieces have been produced).
Before the creation of the algorithm, the programmer must determine a finite number of cases to be resolved.
The expression to be programmed consists in one part which analyzes the value of one or more variables and a second part which determines the action to be taken.
The value of a variable is analyzed by means of comparison operators. The conditional expression and the respective action to be taken are programmed by means of logical operators.

31.2 Comparison operators
These express a comparison meaning between two values.

==	Same.
<>	Different.
>	Higher.
<	Lower.

>=	Higher or same.
<=	Lower or same.

31.3 Logical operators

These are the operators which allow for the definition of the logic of the conditional expression.

IF	Conditional 'If'.
GOTOB	Go back.
GOTOF	Go forward.
GOTO	Go forward and then search for the backward destination.
AND	Associative 'And'.
OR	Or.
IF-ELSE-ENDIF	Control structure with choice between two alternatives. If the condition expressed in the IF block is met (true condition), the following program block is executed until the ELSE function is reached and then continues after the function ENDIF. If the condition expressed in the IF block is not met (false condition), the program part inserted after ELSE is carried out. The second alternative programmed after ELSE can also be left out.
WHILE-ENDWHILE	Control structure with repetition of part of the program. If the condition is met (true condition), the programmed blocks between the functions WHILE and ENDWHILE continue to be executed.

31.4 STOPRE: stop function for the program execution

When the STOPRE function is inserted, the numeric control interrupts the execution of the program until all the previously programmed blocks have been executed.

In programming with variables, this function allows for the completion of calculation execution before the obtained result is used in an expression which is subsequently programmed.

STOPRE guarantees that all which has been programmed up to that point has been executed and saved by the NC.

If this function is not used, the NC can carry out the expression even before the values contained therein are updated.

Programming syntax:
```
R8=(R8+1)
STOPRE
IF (R8<>6) GOTOB PASS
```

31.5 Practical exercise

31.5.1 Creation of the algorithm
The program PRG_31_01 in the folder CHAP_29_31 creates a series of holes on the workpiece circumference.

The programmer has recognized five problems to be resolved in the execution of the program:
1) The angular position of the first hole is variable and the operator must be able to set it.
2) The number of holes is variable and the operator must be able to set it.
3) The holes to be created can be equidistantly distributed over 360 degrees (fig. 203).
4) The holes to be created are equidistant but not distributed over 360 degrees (fig. 204). The angle between the holes must be settable.
5) The position of the holes in Z must be settable.

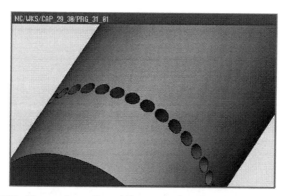

Fig. 203. Execution of equidistant holes distributed over 360°

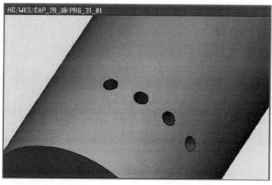

Fig. 204. Execution of a series of holes with constant lead, but not distributed over 360°

Every problem needs to be associated with a variable:
1) R0 = angular position of the first hole.
2) R1 = number of holes to create.
3) R2 = are the holes equidistantly over 360 degrees? (YES=1, NO=0).
4) R3 = angle between the holes when not distributed over 360°
5) R4 = absolute value of the holes on the Z-axis

```
N10 ; workpiece dimensions:
N20 ; XA = 80 bar diameter
N30 ; ZA = 0 machining allowance on front face
N40 ; ZI = -200 length of finished part
N50 ; ZB = -150 extension from jaws
N60
N70 WORKPIECE(,,,"CYLINDER",0,0,-200,-150,80)

; ATTRIBUTION OF THE VALUE TO THE VARIABLES
N80 R0=0 ; WHAT IS THE ANGULAR POSITION OF THE FIRST HOLE
N90 R1=4 ; HOW MANY HOLES DO YOU WANT TO EXECUTE
N100 R2=0 ; ARE THEY EQUIDISTANT OVER 360 DEGREES: 1=YES 0=NO
N110 R4=-40 ; WHAT IS THE ABSOLUTE VALUE IN Z OF THE HOLES

; IF NOT EQUIDISTANT, SET STAGGER ANGLE BETWEEN HOLES
N120 R3=20 ; STAGGER ANGLE BETWEEN HOLES

; AUTOMATIC CALCULATIONS
N130 R8=0 ; HOLE COUNTER RESET
N140 STOPRE

; CALCULATION OF THE STAGGER ANGLE
N150 IF(R2==1) ; IF THE HOLES ARE EQUIDISTANT OVER 360
DEGREES EXECUTE NEXT BLOCK OTHERWISE EXECUTE ELSE
N160 R10=360/R1
N170 ELSE
N180 R10=R3 ; STAGGER ANGLE SET BY OPERATOR
N190 ENDIF

N200 G0 G18 G54 G90
N210 G0 X400 Z500 M8

N220 SETMS(3)
N230 T8 D1; RIGHT HAND RADIAL DRILL D.6
N240 G95 S1200 M3

N250 SPOS[1]=R0
```

```
N260 HOLE1:
N270 G0 Z=R4
N280 G0 X84
N290 G1 X50 F0.1
N300 G0 X84

N310 R8=(R8+1) ; INCREMENTATION OF COUNTER
N320 STOPRE

N330 IF (R8<>R1)
N340 SPOS[1]=(r0+(r10*r8))
N350 GOTOB HOLE1
N360 ENDIF

N370 G0 X200
N380 G0 Z200
N390 M30
```

If the calculated result in block N340 becomes higher than 360°, the numeric control gives the following alarm:

31.5.2 Parametric counter with WHILE

In the program PRG_31_02 contained in the folder CHAP_29_31, 6 holes staggered by 60° are executed by using a parametric counter created with the WHILE cycle.

```
N10 ; workpiece dimensions:
N20 ; XA = 80 bar diameter
N30 ; ZA = 0 machining allowance on front face
N40 ; ZI = -200 length of finished part
N50 ; ZB = -150 extension from jaws
N60
N70 WORKPIECE(,,,"CYLINDER",0,0,-200,-150,80)

N80 R8=0 ; COUNTER RESET

N90 G0 G18 G54 G90
N100 G0 X400 Z500
N110 M8

N120 SETMS(3)
N130 T8 D1; RIGHT HAND RADIAL DRILL D.6
```

```
N140 G95 S1200 M3

N150 SPOS[1]=0
```

N160 WHILE (R8<>6)

```
N170 G0 Z-10
N180 G0 X84
N190 G1 X50 F0.1
N200 G0 X84
N210 SPOS[1]=IC(60)
```
N220 R8=(R8+1) ; INCREMENTATION OF COUNTER
```
N230 STOPRE
```

N240 ENDWHILE

```
N250 G0 X200
N260 G0 Z200
N270 M30
```

31.5.3 Parametric counter with IF

In the program PRG_31_03 contained in the folder CHAP_29_31, 6 holes staggered by 60° are executed by using a parametric counter created with the IF cycle.

```
N10 ; workpiece dimensions:
N20 ; XA = 80 bar diameter
N30 ; ZA = 0 machining allowance on front face
N40 ; ZI = -200 length of finished part
N50 ; ZB = -150 extension from jaws
N60
N70 WORKPIECE(,,,"CYLINDER",0,0,-200,-150,80)

N80 R8=0

N90 G0 G18 G54 G90
N100 G0 X400 Z500
N110 M8

N120 SETMS(3)
N130 T8 D1; RIGHT HAND RADIAL DRILL D.6
N140 G95 S1200 M3

N150 SPOS[1]=0
```
N160 HOLE: ; RETURN LABEL

```
N170 G0 Z-10
N180 G0 X84
N190 G1 X50 F0.1
N200 G0 X84
N210 SPOS[1]=IC(60)
N220 R8=(R8+1) ; INCREMENTATION OF COUNTER
N230 STOPRE

N240 IF(R8<>6) GOTOB HOLE

N250 G0 X200
N260 G0 Z200
N270 M30
```

31.5.4 Algorithm for the execution of 36 longitudinal holes

In the program PRG_31_04 contained in the folder CHAP_29_31, 36 millings distributed equidistantly over 360° are executed; every milling is therefore at 10° from the following one.

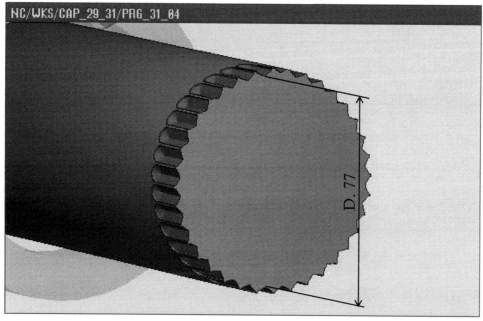

Fig. 205. Part executed with the algorithm entered in the program PRG_31_04

The following program uses the radial drill with diameter 6 mm in position 8 for the simulation of the mill creating the 36 slots. The starting angle is 0°. The first pass is executed at diameter 79 mm. Then 5 more passes are executed, incrementing the cutting depth by 0.2 mm radial (corresponding to 0.4 mm diametral) up to the finished diameter of 77 mm.

```
; workpiece dimensions:
; XA = 80 bar diameter
; ZA = 0 machining allowance on front face
; ZI = -200 length of finished part
; ZB = -150 extension from jaws

N10 WORKPIECE(,,,"CYLINDER",0,0,-200,-150,80)
N20 G0 G18 G54 G90
N30 G0 X400 Z500

N40 SETMS(3)
N50 T8 D1; RIGHT HAND RADIAL DRILL D.6
N60 G95 S4000 M3
N70 G0 Z5
N80 G0 X79 ; POSITIONING AT VALUE OF FIRST PASS

N90 SPOS[1]=0

N120 R8=0 ; ANGULAR COUNTER

N110 START_OP:
N120 R8=0 ; PASS COUNTER
N130 PASS:

N140 G1 G95 Z-16 F0.1
N150 G0 Z5
N160 G0 X=IC(-0.2)

N170 R8=(R8+1)
N180 STOPRE

N190 IF (R8<>6) GOTOB PASS

N200 G0 X79

N210 SPOS[1]=IC(10)
N220 R9=(R9+1)
N230 STOPRE
```

N240 IF (R9<>36) GOTOB START_OP

N250 G0 X200
N260 G0 Z200

N270 M30

32. Three-Axis Mill: Programming Basics

32.1 Introduction
The ISO functions applied to the lathe up until now are the same functions which allow for the programming of a 3-axis mill.
In a lathe, the main working plane is the X-Z plane defined by the function G18.
In a 3-axis mill, the main working plane is the X-Y plane defined by the function G17. On this plane, the tool, rotated by the spindle, moves to execute the profile described in the program, while the position of the tool on the Z-axis determines the execution depth of the tooling operation.

32.2 Layout of the axes in a vertical mill
The scheme for the layout of the axes is the one shown in chapter 4.

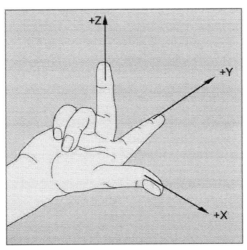

Fig. 206. The same right hand rule applies both to the lathe and to the mill.

The machine under examination is a mill with three axes: X, Y, and Z. The axes X and Y are applied to the machine table while the Z-axis is applied to the tool carrying slide.

Fig. 207. Positive direction of the axes: the arrows show the movements of the tool compared to the workpiece

The positive direction of all the axes are always applied to the tool which moves on the workpiece.

32.3 Machine zero point and definition of the part zero point

While in a lathe the machine zero point is always located on the spindle nose, in a mill it varies from machine to machine according to the manufacturer's choices.

The machine zero point on the X- and Y-axes in some machines is placed on an edge of the table, in other machines it is at the center of the table at the intersection of the diagonals.

The machine zero point on the Z-axis can be positioned on the plane of the table or high up close to stroke limit (as shown in figure 208).

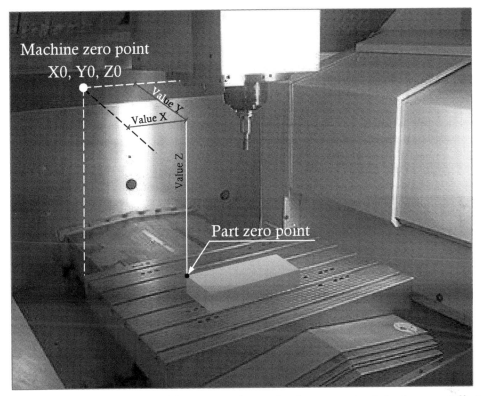

Fig. 208. Position of the machine zero point and values to enter into the zero offset function for the definition of the part zero point

The part zero point is defined by moving the machine zero point on the three axes to the point chosen by the operator. The functions to be used are those for the zero offset (G54, ..., G57) seen in chapter 6.

The values to enter correspond to the distance between the machine zero point and the part zero point as shown in the previous figure.

On the basis of the scheme showing the positive direction of the axes (figure 206), it can be deduced that in this case the value of Z is negative, the value of X is positive and the value of Y is negative.
The manufacturer's manual gives the position of the machine zero point.

32.4 TRANS/ATRANS: incremental shift of the part zero point

The TRANS function allows for the shifting by the program of the part zero point by incrementing the values entered into the absolute zero offset functions: G54, G55, G56 and G57.

The TRANS command must be followed by the letter referring to the axis and by the incremental offset value to execute.

The programming of 'TRANS Y50' means that one wants to increment the active absolute zero offset (e.g. G54) by 50 mm in the positive direction of the Y-axis.

TRANS may be programmed on all the linear axes (X, Y, Z) present in the machine. The ATRANS function further increments the zero offset programmed with the TRANS function.

In a milling machine these functions can be used to replicate the execution of a program part at various points of the workpiece (see figure 209-2).

Another option is to mount various blank parts on the machine table and shift the complete execution of the program from one blank part to the other.

In a lathe these functions are more commonly used when it is decided to define the part zero point within the program. It is for example possible to use the function G54 to shift the machine zero point to a fixed point (e.g. the face of the jaws) to then program the TRANS function, followed by the letter 'Z' associated with the extension of the workpiece face from the jaws (see figure 209-1).

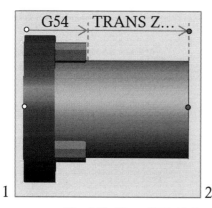

Fig. 209. Use of TRANS: 1: in a lathe; 2: in a milling machine

32.5 Position of the point controlled by the NC and tool geometry

The point controlled by the NC is always placed by the manufacturer at the spindle's center of rotation on the tool holder plane of attachment (fig. 210-1). The tool attachments are normally standardized and have a conical shape, of varying dimension according to the maximum dimension of the tools which can be mounted to the machine.

The rotating centre of all the tools used in milling machine is concentric to the point moved by the NC. This means that the geometry values on the X- and Y-axes are always equal to zero, while the zero offset value on the Z-axis corresponds to the distance between the tip of the tool and the attachment plane of the tool holder (fig. 210-2).

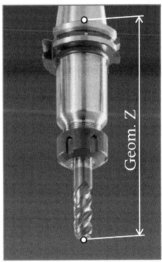

Fig. 210. 1:Point moved by the NC; 2:Offset value of a mill on the Z-axis

To obtain the offset value in Z by touching the workpiece:
- you program the activation of the part zero point in MDA by using the zero offset function set in the program (G54),
- you touch the workpiece at a known value with respect to the part zero point,
- you enter this value and you activate the automatic calculation.

Another method is to measure the tool geometry outside of the machine either using an altimeter or by means of specific external measuring system as the one shown in figure 211.

The systems for external presetting have a tool attachment which is identical to the one present in the machine. Before proceeding to the measuring of the tool length, the tool holder is offset by making the measured zero coincide with the point moved by the NC (in this case on the face of the attachment cone); then the tool is fixed to the support in order to measure its length and, if necessary, other characteristic elements like the diameter or the length of the cutting edge.

A camera with video helps the operator to define exactly the measuring point.

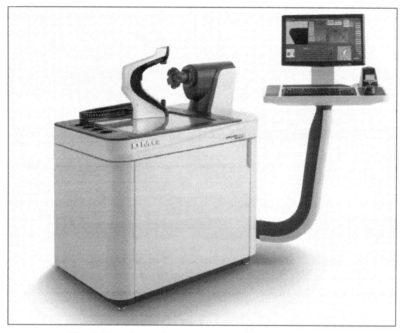

Fig. 211. Measuring system outside the machine

The diameter of the mill (inserted in the geometry page) will be considered by the machine by programming of the tool radius compensation functions.

Contrary to lathes, in mills the concept of tool quadrant code does not exist, as all the tools used in those machines are offset at the center and are therefore always defined by quadrant zero. **To review the tool radius compensation functions and the quadrant codes, please refer to chapter 15.**

32.6 Setting of tool rotation and feedrate

In chapter 8, the functions for the activation of the spindle rotation in a lathe have been described. In a mill, this is much more simple.

In the lathe, the setting of the spindle rotation with fixed number of revolutions or with constant cutting speed was examined; in the mill, the only alternative is that of a fixed number of revolutions calculated on the basis of the tool diameter used.

The number of revolutions of the mill is calculated using the formula shown in figure 75.

In the case of a mill with a diameter of 32 mm, which works at a cutting speed of 100 m/min., the number of revolutions to be programmed corresponds to:

$$n = (1000 \times 100) / (32 \times 3.14) = 995 \text{ rev./min.}$$

In chapter 9, the functions for the setting of the feedrate were examined. In this case the functions G95, G94 and all the concepts associated with them are applicable to the mill exactly as already seen for the lathe.

32.7 Practical exercise

32.7.1 Introduction
As we have arrived at the end of the course and explained the use of the ISO functions, we will now see a complete programming example for a part created on a 3-axis mill.
The drawing shown in figure 221 represents the part to be created.

32.7.2 Creation of a three-axis mill (X, Y, Z)
Before proceeding, it is necessary to create the mill to use in this chapter in the training and simulation software.
Once you've started the *SinuTrain Operate* Program, push the NEW icon in the upper part of the screen, leave selected the first point *Create a new machine configuration from a template* and push NEXT.
This window shows you a list of standard machine templates preconfigured in SinuTrain, select *Simple vertical milling machine* and press NEXT.
Now choose the name of the milling machine, describe its basic features, set the language you want to use and the dimensions of the window which plays the video of the machine according to the following information:

GENERAL	*Machine name:* MILLING MACHINE: 3 axes *Description:* SP1-spindle (main spindle), X-axis (linear geometry axis), Y-axis (linear geometry axis), Z-axis (linear geometry axis),
LANGUAGE	English - English
RESOLUTION	640x480

Push FINISH. The machine has been created and is now shown on the starting page of the program.

Select the newly created mill and push the START icon.

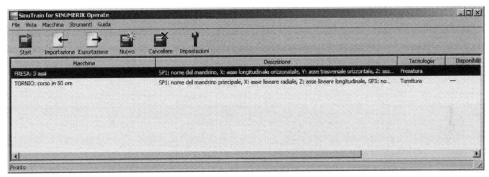

Fig. 212. Start-up of the mill in the training software

32.7.3 Download of the programs and import into SinuTrain
Open the website www.cncwebschool.com and access the DOWNLOAD area, click on MILL_3_AXES and download the folder T3_PROG containing the programs used in this chapter.
Select the downloaded compressed folder with the pointer, push the right button of your mouse and select: *Extract all, Next, Next*.

Now import the programs into the training software and close the SinuTrain program selecting *File* end then *Exit*.

Copy the folder M3_PROG onto an empty USB stick.

Open SinuTrain again, select the newly created mill and push START from the icons in the upper part of the screen.
On the control panel, click PROGRAM MANAGER.
After selecting USB from the horizontal softkeys, the content of the USB memory is shown.

Select the folder M3_PROG with the arrows and push the yellow INPUT button to open it.

318

Place the orange selection on the folder CHAP_32.
Press the vertical softkey COPY.
Push NC from the horizontal softkeys.
Move down with the arrows until you have selected the folder WORKPIECES, the press PASTE from the vertical softkeys.
Open the WORKPIECE folder with INPUT, open the folder CHAP_32 with INPUT and, also with INPUT, open the test program PRG_0.

32.7.4 Direct selection of the tools in the program
The creation of new tools and the export and import of tooling data follow the same rules as seen in chapter 7.
For this exercise it is not necessary to set any new tools as the program uses the tools which are already present in the machine.

It is furthermore possible to use a new method to select the tools in the program: instead of programming the position T and the geometry D, the tool to be used is chosen directly from the magazine.

Position the cursor on the block from which to call the tool and press the vertical button TOOL SELECTION.

Fig. 213. Page for the selection of the tools directly from the magazine

A list of available tools is shown.
With the arrows, highlight the tool to be called and confirm your selection with OK. The name of the tool is inserted in the program.

As will be seen in the example program, **the tool selection needs to be completed by adding the function D1 to the same block to call the offset values and the function M6 to enable the tool changing procedure**.
The block must be programmed as follows:

T="CUTTER 16" D1 M6

To select the tool offset values see paragraph 7.3, for the function M6 see figure 45 and paragraph 7.2.

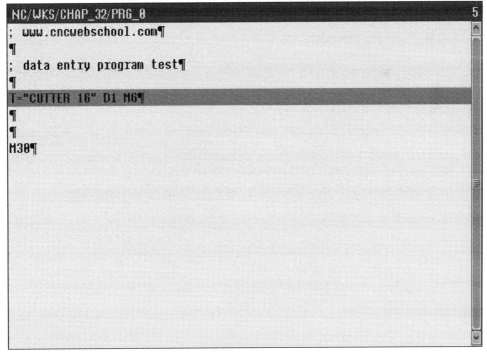

Fig. 214. Completion of the direct tool call instruction in the program

32.7.5 Graphic definition of the blank part

Now close the program PRG_0.
Open the program PRG_32_01 and open, at block N10, the dialog box for the insertion of the blank part data.

N10 WORKPIECE(,,,"RECTANGLE",0,0,-32,-150,160,120)

By means of the graphic definition of the blank part the following elements are defined:
- shape of the blank part,
- position on X and Y of the part zero point,
- position on Z of the part zero point,
- dimensions of the blank part.

Below is the explanation of the parameters based on the different options that can be selected.

The selection CYLINDER forces the position of the part zero point in X and Y onto the main axis of the cylinder.

Blank part:	Cylinder
XA:	Cylinder diameter.
ZA:	Position of the upper face of the workpiece referring to the part zero point.
ZI - **absolute**: - **incremental**:	Distance from the lower face of the workpiece: **referring to the part zero point.** **referring to the upper face.**

Fig. 215. Description of the blank part dimensions: CYLINDER

Below are some examples applied to a 32 mm high blank part which help understand how the graphic simulation interprets the values associated with ZA and ZI.

If you want to simulate the position of the **part zero point on the upper face of the blank part,** set the following values:

ZA = 0:   shows that the upper face of the blank part corresponds to the position of the part zero point.

ZI = -32   (if the value is expressed in **absolute** coordinates) shows that the lower face of the blank part is located at 32 mm in the negative direction **compared to the part zero point**.

ZI = -32   (if the value is expressed in **incremental** coordinates) shows that the lower face of the blank part is located at a distance of 32 mm in the negative direction **compared to the upper face**.

If ZA equals zero, the position of the part zero point corresponds to the upper face of the blank part; therefore the value ZB which expresses the total height of the blank part (defining the position of its lower face) is equal both when expressed in absolute coordinates and when expressed in incremental coordinates.

If you want to simulate the position of the **part zero point on the machine table,** set the following values:

ZA = 32:   shows that the upper face of the blank part is located at a distance of 32 mm in the positive direction of the part zero point.

ZI = 0   (if the value is expressed in **absolute** coordinates) shows that the lower face of the blank part corresponds to the position of the part zero point.

ZI = -32   (if the value is expressed in **incremental** coordinates) shows that the lower face of the blank part is located at a distance of 32 mm in the negative direction compared to the upper face.

The selection PIPE forces the position of the part zero point in X and Y onto its main axis.

Blank part:	Pipe
XA:	External diameter of the tube.
XI:	Internal diameter of the tube.

Fig. 216. Description of the blank part dimensions: PIPE

The selection BLOCK CENTERED forces the position of the part zero point in X and Z onto the intersection of the rectangle's diagonals.

Blank part:	Block Centered
W:	Side of the rectangle positioned along the Y-axis.
L:	Side of the rectangle positioned along the X-axis.

Fig. 217. Description of the blank part dimensions: BLOCK CENTERED

The selection BLOCK refers the basic position of the part zero point in X and Y to the edge at the lower left side of the rectangle.

Blank part:	Block
X0:	Coordinate X of the edge referring to the part zero point.
Y0:	Coordinate Y of the edge referring to the part zero point.
X1:	Coordinate X of the opposite edge referring to the part zero point (abs.) or to the first edge (incr.).
Y1:	Coordinate Y of the opposite edge referring to the part zero point (abs.) or to the first edge (incr.).

Fig. 218. Description of the workpiece dimensions: BLOCK

In the case of the parallelepiped, in order to bring the part zero point to the center of the diagonals of a rectangle with 160 mm side on the X-axis and of 120 mm side on the Y-axis, set the data as shown in the following figure:

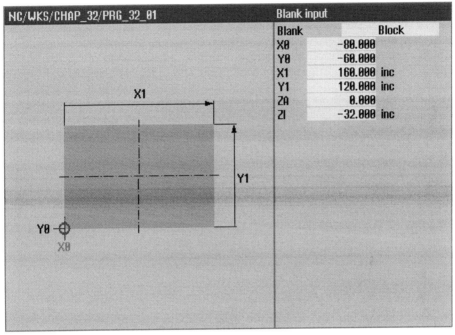

Fig. 219. Description of the blank part dimensions: BLOCK

The selection N CORNER forces the position of the part zero point in X and Y onto the intersection of the polygon's diagonals.

Workpiece:	N Corner
N:	Number of edges of the polygon.
SW:	Dimension of the polygon's key (available only for polygons with even number of edges).

Fig. 220. Description of the workpiece dimensions: N CORNER

32.7.6 Drawing of the part to be created

The part zero point is located on the intersection of the diagonals on the upper face of the workpiece as specified in the drawing.

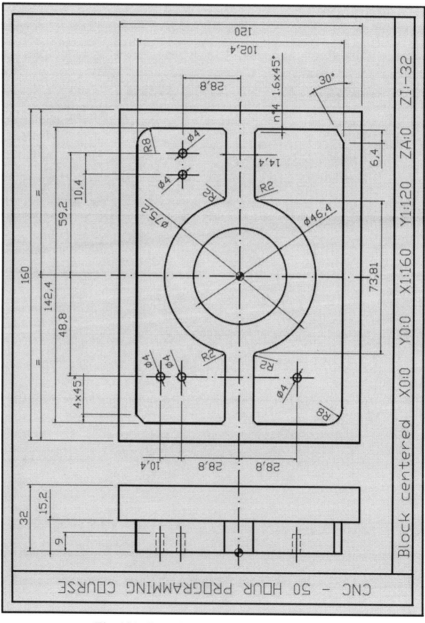

Fig. 221. Drawing of the part to be created

32.7.7 Program, phase 1: execution of the external profile

Activate the graphic simulation of the program PRG32_01 and associate the tool movements with the blocks programmed hereafter. Chamfers, rounds and angles are defined in the tool path by means of the direct programming functions explained in chapter 12.

```
N10  WORKPIECE(,,,"RECTANGLE",0,0,-32,-150,160,120)
N20  G17 G54 G90
N30  G0 Z500
N40  T="CUTTER 16" D1 M6 ; MILL DIAM. 16
N50  G95 S2800 M3 M8 F0.2
N60  G0 Y0 X90
N70  G0 Z-15.2
N80  G1 X71.2 G41
N90  Y-48
N100 Y-51.4 ANG=210
N110 X-71.2 RND=8
N120 Y51.2 CHF=4
N130 X71.2 RND=8
N140 Y-2
N150 X82 G40
```

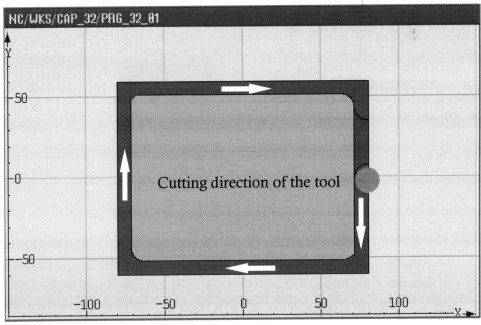

Fig. 222. Creation of the external profile

32.7.8 Program, phase 2: roughing of the internal profile

The programming of the circular interpolations is carried out according to the instructions in chapter 13.

```
N170 T="CUTTER 10" D1 M6 ; MILL DIAM. 10
N180 G95 S2500 M3 M8 F0.16
N190 G0 Y0 X80
N200 G0 Z-15.2
N210 G1 X30.4
N220 G3 X30.4 Y0 I-30.4
N230 G3 X-30.4 Y0 CR=30.4 F1
N240 G1 X-74

N250 G0 Z500
```

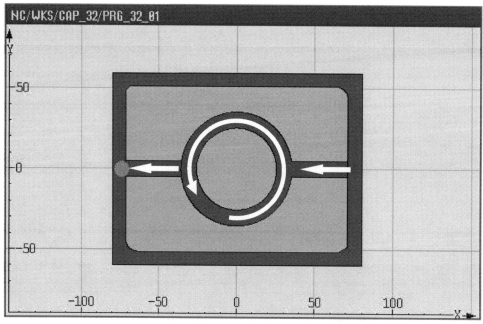

Fig. 223. Roughing of the internal profile

NOTE: the letters I, K and J stand for the coordinates of the radius center referring to the starting point of the arc on the X, Z, and Y axis respectively.

32.7.9 Program, phase 3: finishing of the internal profile

In this program part the internal profile of the workpiece is finished.

```
N260 T="CUTTER 10" D1 M6 ; MILL DIAM. 10
N270 G95 S3200 M3 M8 F0.16
N280 G0 Y12 X80
N290 G0 Z-15.2
N300 G1 X71.2 G41
N310 G1 Y7.2 CHF=1.6
N320 G1 X36.905 RND=2
N330 G3 X-36.905 Y7.2 CR=37.6 RND=2
N340 G1 X-71.2 CHF=1.6
N350 G1 Y12
N360 G1 X-80 G40

N370 G1 Y-12 X-80 F1
N380 G1 X-71.2 G41 F0.16
N390 G1 Y-7.2 CHF=1.6
N400 G1 X-36.905 RND=2
N410 G3 X36.905 Y-7.2 CR=37.6 RND=2
N420 G1 X71.2 CHF=1.6
N430 G1 Y-12
N440 G1 X80 G40

N450 G0 Y0
N460 G0 X32
N470 G1 X23.2 G41
N480 G2 X23.2 Y0 I=-23.2
N490 G1 X32 G40

N500 G0 Z500
```

At block N260, the finishing of the upper internal profile begins.

At block N370, the finishing of the lower internal profile begins.

At block N450, the finishing of the internal diameter of 46.4 mm begins.

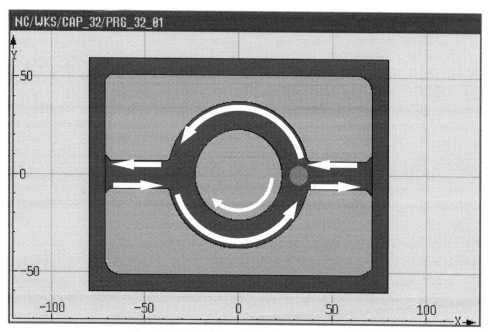

Fig. 224. Finishing of the internal profile

32.7.10 Program, phase 4: execution of the holes

The holes are programmed by using the drilling cycle and the MCALL function as explained in chapter 23.

```
N510 T="CUTTER 4" D1 M6 ; MILL DIAM. 4
N520 G95 S2300 M3 M8 F0.12
N530 G0 X59.2 Y28.8
N540 G0 Z2

N550 MCALL CYCLE82(10,0,2,-9,,0.6,0,1,12)

N560 G0 X59.2 Y28.8
N570 G0 X48.8 Y28.8
N580 G0 X-48.8 Y39.2
N590 G0 X-48.8 Y28.8
N600 G0 X-48.8 Y-28.8
N610 MCALL

N620 G0 Z500
```

329

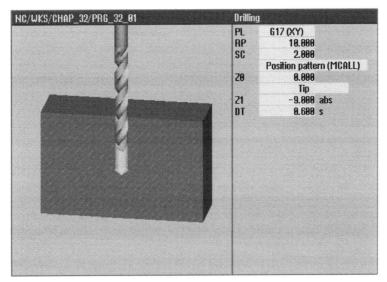

Fig. 225. Data entered in drilling cycle

32.7.11 Program, phase 5: activation of graphic simulation

Open the program PRG_32_01 in the folder CHAP_32 and activate the graphic simulation. Then execute the program in single block mode and reduce the execution speed by setting the potentiometer in the graphic display to 80%.

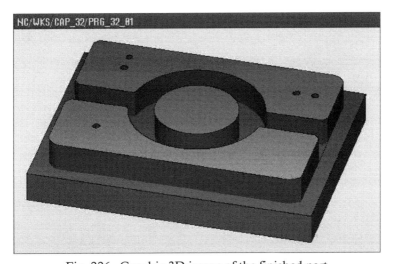

Fig. 226. Graphic 3D image of the finished part

32.7.12 Programming example with the use of TRANS

The program PRG_32_02 in the folder CHAP_32 creates the part shown in the following figure. It is characterized by the presence of four identical tooling operations executed in four different positions.

The part zero point is located at the center of the rectangle; the program uses the TRANS function to shift it on the X- and Y-axes to the four points from which to start the execution of the profile and the holes.

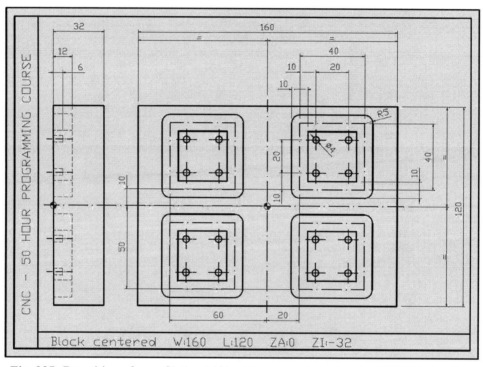

Fig. 227. Repetition of a profile by shifting the zero point using the TRANS function

```
; blank part: centered parallelepiped
; W = 120 side length on Y
; L = 160 side length on X
; ZA = 0 position of upper face with respect to part zero point
; ZI = -32 position of lower face with respect to part zero point

N10 WORKPIECE(,,,"RECTANGLE",64,0,-32,-150,160,120)
N20 G17 G54 G90
N30 G0 Z500
```

```
N40 T="CUTTER 10" D1 M6 ; MILL DIAM. 10
N50 G95 S2800 M3 M8
```

N60 TRANS X20 Y10

```
N70 PROFILE1:
N80 G0 Y0 X0
N90 G0 Z2
N100 G1 Z-12 F0.1
N110 G1 X40 F0.18
N120 G1 Y40
N130 G1 X0
N140 G1 Y0
N150 G1 Z5 F0.8
N160 END1:
```

N170 TRANS X-60 Y10

```
N180 REPEAT PROFILE1 END1
```

N190 TRANS X-60 Y-50

```
N200 REPEAT PROFILE1 END1
```

N210 TRANS X20 Y-50

```
N220 REPEAT PROFILE1 END1

N230 G0 Z500


N240 T="CUTTER 4" D1 M6 ; MILL DIAM. 4
N250 G95 S2300 M3 M8 F0.12
```

N260 TRANS X20 Y10

```
N270 DRILLING1:
N280 G0 X10 Y10
N290 G0 Z2
N300 MCALL CYCLE82(5,0,2,-6,,0.6,0,1,12)
N310 G0 X10 Y10
N320 G0 X30 Y10
N330 G0 X30 Y30
N340 G0 X10 Y30
N350 MCALL
N360 END_DRILLING1:
```

N370 TRANS X-60 Y10

```
N380 REPEAT DRILLING1 END_DRILLING1
```

N390 TRANS X-60 Y-50
N400 REPEAT DRILLING1 END_DRILLING1

N410 TRANS X20 Y-50
N420 REPEAT DRILLING END_DRILLING1

N430 G0 Z500

N440 M30

33. Translation of technical terms

33.1 Machine components

English	Italiano	Deutsch
Electrical cabinet	Armadio elettrico	Schaltschrank
Chuck	Autocentrante	Spannfutter
Base, machine bed	Basamento	Maschinenbett
Magazine, loader	Caricatore di barre	Stangenlademagazin
Cable, wire	Cavo	Kabel
Belt	Cinghia	Riemen
Electrical plug	Connettore, spina	Steckverbinder
Counterspindle	Contromandrino	Gegenspindel
Tail stock	Contropunta	Reitstock
Bearing	Cuscinetto	Kugellager
Filter	Filtro	Filter
Flow switch	Flussostato	Durchflussregler
Joint	Giunto	Kupplung
Jaws	Griffe	Spannbacken
Guide	Guida	Führung
Main switch	Interruttore principale	Hauptschalter
Spindle	Mandrino	Spindel
Pressure gauge	Manometro	Druckmesser
Engine, motor	Motore	Motor
Lubricating oil	Olio lubrificante	Schmieröl
Coolant	Olio refrigerante	Schneidöl
Control panel	Pannello di controllo	Bedienpult

Sliding block	Pattino	Gleitbacke
Collet	Pinza elastica	Spannzange
Tool holder	Portautensile	Werkzeughalter
Potentiometer	Potenziometro	Potentiometer
Pulley	Puleggia	Scheibe
Gantry	Robot a portale	Portal
Chips conveyor	Scaricatore di trucioli	Späneförderer
Screen, video	Schermo	Bildschirm
Sensor	Sensore	Sensor
Slide	Slitta	Schlitten
Buttons, keys	Tasti, pulsanti	Taste
Lathe	Tornio	Drehmaschine
Turret	Torretta rotante	Werkzeugrevolver
Ball screw	Vite a ricircolo di sfere	Kugelumlaufspindel
Handle	Volantino	Handrad

33.2 Programming

English	Italiano	Deutsch
Algorithm	Algoritmo	Algorithmus
Disengagement	Allontanamento	Wegfahren
Angle	Angolo	Winkel
Arc	Arco	Bogen
Rotating axis	Asse di rotazione	Drehmitte
Main axis	Asse principale	Hauptachse
Activation, enabling	Attivazione	Aktivierung
Feedrate	Avanzamento	Vorschub
Approaching	Avvicinamento	Anfahren
Tool offset	Azzeramento utensile	Werkzeug setzen
Center	Centro	Zentrum, Mitte
Circle	Cerchio	Kreis

Tool radius compensation	Compensazione raggio utensile	Werkzeugradius-kompensation
Absolute coordinates	Coordinate assolute	Absolute Koordinaten
Incremental coordinates	Coordinate incrementali	Inkremetale Koordinaten
Tool data	Dati utensile	Werkzeugdaten
Direction	Direzione	Richtung
Disable	Disabilitare	Deaktivieren
Function	Funzione	Funktion
Interpolation	Interpolazione	Interpolation
Tooling operation	Lavorazione con utensile	Werkzeugbearbeitung
Line	Linea	Linie
Length	Lunghezza	Länge
Movement	Movimento	Bewegung
Number of revolutions	Numero di giri	Drehzahl
Part, workpiece	Pezzo	Stück, Werkstück
Working plane	Piano di lavoro	Arbeitsplan
Program	Programma	Programm
Main program	Programma principale	Hauptprogramm
Radius	Raccordo	Radius
Rapid	Rapido	Eilgang
Reference	Riferimento	Referenz
Rotation	Rotazione	Rotation, Schwenkung
Conditional skips	Salti condizionati	Bedingte Sprünge
Counterclockwise	Senso antiorario	Gegenuhrzeigersinn
Clockwise	Senso orario	Uhrzeigersinn
Graphic simulation	Simulazione grafica	Grafische Simulation
Coordinate system	Sistema di riferimento	Referenzsystem
Chamfer	Smusso	Fase
Subprogram	Sottoprogramma	Unterprogramm
Dwell time	Tempo di attesa	Verweilzeit

Tool	Utensile	Werkzeug
Variable	Variable	Variable
Cutting speed	Cutting speed	Schnittgeschwindigkeit
Machine zero point	Zero macchina	Maschinennullpunkt
Part zero point	Zero pezzo	Werkstücknullpunkt

33.3 Tooling operations

English	Italiano	Deutsch
Reamer	Alesatore	Reibahle
Boring	Barenatura	Ausbohren
Boring bar	Bareno	Bohrstange
Gear hobbing	Dentatura di ingranaggi	Verzahnung
Threading	Filettatura	Gewindeschneiden
Thread	Filetto	Gewinde
Finishing	Finitura	Schlichten
Deep drilling	Foratura profonda	Tieflochbohren
Radial drilling	Foratura radiale	Querbohren
Hole	Foro	Loch, Bohrung
End mill	Fresa a candela	Nutenfräse
Disk mill	Fresa a disco	Scheibenfräse
Milling	Fresatura	Fräsen
Knurling	Godronatura	Rändelung
Groove	Gola	Nut
Compensated tapping operation	Maschiatura compensata	Gewindebohren mit Ausgleichsfutter
Rigid tapping	Maschiatura rigida	Gewindebohren ohne Ausgleichsfutter
Grinding	Operazione di rettifica	Schleifen
Drill	Punta	Bohrer
Undercut	Scarico	Freistich

Facing	Sfacciatura	Plandrehen
Roughing	Sgrossatura	Schruppen
Parting off	Troncatura	Abstechen
Tapping tool	Ut. per maschiare	Gewindebohrer

33.4 Materials and related terms

English	Italiano	Deutsch
Steel	Acciaio	Stahl
Stainless steel	Acciaio inossidabile	Edelstahl
Aluminium	Alluminio	Aluminium
Silver	Argento	Silber
Bar	Barra	Stange
Cylindrical bar	Barra cilindrica	Zylindrische Stange
Hexagonal bar	Barra esagonale	Sechskantstange
Bronze	Bronzo	Bronze
Iron	Ferro	Eisen
Melting	Fusione	Fusion
Cast iron	Ghisa	Gusseisen
Gold	Oro	Gold
Brass	Ottone	Messing
Preformed part	Pezzo preformato	Vorgeformtes Teil
Lead	Piombo	Blei
Plastic	Plastica	Kunststoff
Copper	Rame	Kupfer
Tin	Stagno	Zinn
Molded part	Stampato	Gussform
Titanium	Titanio	Titan
Chip	Truciolo	Span
Zinc	Zinco	Zink

33.5 Notes

English	Italiano	Deutsch

Conclusion

*'With commitment, time and method,
you can achieve all your goals'*

As this simple phrase has introduced us to the course, in the same way it marks its conclusion; it expresses the principles which have led me while writing this book and you in your learning process.

Your time spent learning, together with the didactic method provided, has allowed you to achieve an important goal: to gain in-depth knowledge of the programming concepts applied to lathes and mills commanded by a numeric control.

The training software was essential to simulate the working experience on a real machine tool. Now you only have to consolidate the concepts learned by continuous practice.

I hope that the commitment and the passion which led me to write this book have produced a valid instrument able to contribute to your professional growth.

Lorenzo Rausa

References of the figures

The following figures refer to machines built by DMG Group.

Fig. 1, 2, 24, 25, 26, 28, 29, 49, 50, 51, 54, 207, 208, 210, 211.

The following figures are taken from the training course SANDVIK COROMANT.

Fig. 71, 72, 73, 75, 76, 77.

Made in the USA
Middletown, DE
10 May 2017